A Handbook of
Mexican Roadside Flora

CHARLES T. MASON JR.
and
PATRICIA B. MASON

The University of Arizona Press / Tucson

About the Authors

Charles T. Mason Jr., a professor of plant taxonomy, has been a member of the University of Arizona faculty since 1953. As curator of the University Herbarium, identifying plants for a curious public has been among his chief concerns. He has traveled widely in Mexico in search of plants for use in cancer research and many other areas of scholarly inquiry.

Patricia B. Mason received her degrees in painting at the University of California at Los Angeles and Berkeley. A prolific scientific illustrator, her work has enlivened and informed the pages of numerous scientific journals and several books.

THE UNIVERSITY OF ARIZONA PRESS
Copyright © 1987
The Arizona Board of Regents
All Rights Reserved
Manufactured in the U.S.A.
This book was set in 10/12 Linotronic 300 Times Roman.

Library of Congress Cataloging-in-Publication Data

Mason, Charles T.
 A handbook of Mexican roadside flora.

 Bibliography: p.
 Includes index.
 1. Roadside flora—Mexico—Identification.
I. Mason, Patricia B. II. Title.
QK211.M37 1987   581.972   87-10890
ISBN 0-8165-0997-2 (alk. paper)

British Library Cataloguing in Publication data are available.

Contents

Acknowledgments iv

Introduction 1

    Geography and Vegetation 3

    About the Entries 7

Plant Identification Key 9

The Species 23

Glossary 361

Selected References 371

Index 373

Acknowledgments

To our many friends, whose encouragement and suggestions spurred us forward, and to many unknown friends, whose questions showed us a need for our guidebook, we express our sincere appreciation. Special thanks go to Mrs. Rebecca Van Devender and Greg Starr, staff members of the University of Arizona Herbarium, who were particularly helpful in checking details, making recommendations, and discussing problems. Dr. George Cummins critically read an early stage of the manuscript and made many greatly appreciated suggestions. Dr. Edgar Heylmun proposed numerous plants from the Guadalajara region for consideration. Unfortunately, limited space precluded the inclusion of all the suggestions we received.

Introduction

"What is the shrub with showy yellow flowers along the road near Hermosillo?"

"We saw this ball attached to a tree. What is it?"

These and many similar questions brought to the herbarium at the University of Arizona in Tucson by travelers returning from Mexico inspired the writing of *Mexican Roadside Flora*. As it is impossible to present in any meaningful manner the vast number of plants growing in Mexico, an attempt has been made to select those which, because of their flowers, the shape of their fruit, their unusual leaves, or overall peculiarities, have caused us in our many trips to Mexico to say, "Oh! What is that?" It is our intent to include those plants that can be seen from the main highways or, in a few instances, those prominently cultivated around hotels or motels. Most of the plants discussed are trees or shrubs, because these are present all year. There are many showy, unusual annuals, but few have been included, owing to their ephemeral and seasonal nature and vast number.

GEOGRAPHY AND VEGETATION

Each species of plant, while tolerant of variations in ecological factors, grows best where its individual requirements of temperature, moisture, light, and soil quality are met. Several species may have similar or overlapping tolerances, enabling them to grow together. These groups of species are what the botanist may view as a vegetation type, and to which names are applied based upon a particular dominant genus (Pine Forest, for example) or upon a particular life form (Deciduous Tropical Forest). The highly diverse topography of Mexico has produced numerous ecological areas conducive to the development of several vegetation types, each composed of many different species.

INTRODUCTION

When traveling through Mexico, one encounters topographic variations from low, flat coastal plains with a warm climate and high humidity to high volcanic peaks supporting perpetual snow. Large areas shielded from moist oceanic winds result in regions of low rainfall and desert vegetation of succulents and drought–tolerant species. Mountain ranges rising from coastal plains or desert regions support temperate broadleaf forests of oaks or, at higher elevations, evergreen forests of pines. Southern lowlands may be regions of high rainfall, in which one finds forests of large broadleaf evergreen trees and their accompanying epiphytic orchids, bromeliads, and climbing vines.

Approximately half of Mexico's 2,000,000 square kilometers (772,200 square miles) lies on either side of the Tropic of Cancer. In the north, two massive mountain ranges — the Sierra Madre Oriental on the east and the Sierra Madre Occidental on the west — drain oceanic winds of their moisture, resulting in a central portion characterized by low rainfall and desert vegetation. This area is the Altiplano, or Central Plateau, a large flattish plain with small scattered mountain ranges, the whole sloping gently upward from its northern boundary with the United States to higher country near south central Jalisco and northern Hidalgo.

Extending from the water to the mountains on both coasts are the low-lying Coastal Plains. In some regions these are broad and flat, supporting extensive agriculture; at other points the mountains almost form the seashore.

The southern end of the Altiplano is marked by the Volcanic Cordillera, a vast east-west zone of recent volcanic activity. In this transvolcanic area are the highest and best known of Mexico's volcanic peaks: Pico de Orizaba, 5650 m (18,532 ft); Popocatépetl, 5450 m (17,876 ft); Ixtaccíhuatl, 5280 m (17,318 ft); Nevado de Toluca, 4560 m (14,957 ft).

South of the Volcanic Cordillera, and essentially a continuation of the northern mountain chains, are the Sierra Madre de Oaxaca on the east and Sierra Madre del Sur on the west. These end at the very narrow and low-lying Isthmus of Tehuantepec, a region wherein the eastern and western coastal floras intermingle. Enclosed by these mountain ranges is the basin of the Rio Balsas.

Between the Isthmus of Tehuantepec and the Guatemala border, principally in the state of Chiapas, are the Chiapas Highlands and the

INTRODUCTION

Sierra Madre de Chiapas, the latter a volcanic chain connecting with the similar mountains of Guatemala. To the north and east of Chiapas is the low-lying, relatively flat Yucatán Peninsula, a large limestone shield with a thin layer of soil.

Mexico's wide variation in topography, together with associated variation in rainfall, temperature, and soil quality, has enabled several vegetation types to develop. No sharp line delimits the boundary of each assemblage, and some species may be associated with more than one group of species. Although vegetation type is not used as an aid to recognition of species in this book, it might be of interest to know the types of plants to be expected in each assemblage.

Prominent among the dominant aquatic vegetation tolerant of salt or brackish water along the shore edge of the Coastal Plain are the mangroves: *Rhizophora mangle*, *Avicennia germinans*, *Laguncularia racemosa*, and *Conocarpus erecta*. Other aquatic vegetation types will develop in fresh water along streams, lakes, and marshes.

Inland along the Coastal Plain, in areas of low rainfall but high humidity, can be found the Short Tree Forest. This formation develops on the western Coastal Plain from Sonora to the middle of Sinaloa and in isolated areas south to Tehuantepec. On the Atlantic side, the Short Tree Forest can be seen in large areas of the northeast Coastal Plain, including parts of San Luis Potosí and northern Veracruz. It occurs again in the northern part of the Yucatán Peninsula, on the Altiplano in Guanajuato, and in the adjacent states of Michoacan and Queretaro. Approximately 5 percent of Mexico originally was occupied by the Short Tree Forest, but much of the native forest was destroyed as agriculture developed. Some of the dominant plants of the area are *Prosopis juliflora*, *Prosopis velutina*, *Acacia cochliacantha*, *Cercidium praecox*, *Olneya tesota*, *Acacia greggii*, *Haematoxylon brasiletto*, *Lysiloma divaricata*, and *Fouquieria macdougallii*.

In drier areas of the Coastal Plain the trees give way to the Shrubby Desert formation, or Desert Vegetation. This formation develops on the northeast Coastal Plain from Coahuila to Tamaulipas and on the Coastal Plain and low mountains of northern Sonora. It is the most common vegetation type of Baja California. Desert Vegetation also occurs in parts of Puebla and adjacent Oaxaca, and on the Altiplano from Chihuahua and Coahuila south to Jalisco, Guanajuato, Hidalgo, and the State

INTRODUCTION

of México. Some desert species illustrated here are *Agave lechuguilla*, *Euphorbia antisyphilitica*, *Flourensia cernua*, *Jatropha dioica*, *Larrea tridentata*, *Parthenium argentatum*, and *Yucca carnerosana*.

As one leaves the Coastal Plain and climbs to higher elevations, the vegetation gradually changes; trees of the genus *Quercus* appear. The Oak Forest becomes the dominant vegetation in the Sierra Madre Oriental and is very common in the Sierra Madre Occidental between 1200 and 2800 m (4000 to 9000 ft) elevation. Oaks are predominant in the Volcanic Cordillera and have been reported from all states, except Yucatán and Quintana Roo. In addition to several types of oaks, *Arbutus xalapensis* may be found in the drier Oak Forests; in those forests with more moisture, *Oreopanax peltatum* may occur.

With further increase in elevation, pine trees begin to appear; they may at first intermingle with oaks in a Pine-Oak Forest, but eventually they form pure stands of Pine Forest. Species of *Pinus* do best between 1500 and 3000 m (5000 and 10,000 ft), but in some areas they may ascend to elevations of 4100 m (13,500 ft).

The Deciduous Tropical Forest develops along the lower slopes of the Sierra Madre Occidental and above the Coastal Plain from Sinaloa and southwestern Chihuahua south to Chiapas. This vegetation type, characterized by extended periods of leaflessness, also is found in isolated patches along the east coast to the northern part of Yucatán and adjacent Campeche, where it occupies a large area. The Deciduous Tropical Forests develop in areas where the minimum temperature rarely reaches freezing and annual precipitation, although in some areas abundant, is unequally distributed throughout the year. During a dry period lasting five to eight consecutive months, the trees lose their leaves; the area has been described as having a sad, gray, and desolate aspect in drought. Toward the end of the drought its appearance changes and many of the components of this vegetation type burst into flower. Owing to the spectacle of bare trees ablaze with flowers, many of our plant examples come from this group. Among the showy representatives are *Bombax palmeri*, *Ceiba aesculifolia*, *Tabebuia chrysantha*, *T. palmeri*, and *Plumeria acutifolia*.

Where neither precipitation nor minimum temperature is a limiting factor, many large evergreen trees, commonly supporting epiphytes and

INTRODUCTION

vines, develop into the Tropical Evergreen Forests. This vegetation type occurs in southern Tamaulipas and in parts of Veracruz, Tabasco, Chiapas, Oaxaca, and Quintana Roo. Among the representatives of this forest are *Brosimum alicastrum*, *Cedrela odorata*, and *Manilkara achras*.

ABOUT THE ENTRIES

The scientific name of a plant consists of two words, as in *Agave americana*; *Agave* is the generic name, *americana* the specific epithet. Both are Latinized words, and the latter may be descriptive or commemorative; as a descriptive term it tells something about the plant. *Americana* indicates that our example originally was thought to be native to America. Frequently the specific epithet is used to commemorate the name of the person who first found the species.

The species is the smallest of the taxonomic groups with which we are concerned. It is composed of those individuals that have certain fundamental characteristics, primarily flower characters, that are alike. Just as people vary in size, shape, and color yet represent one species, so will plants belonging to one species vary.

The genus is the taxonomic group that includes one or more species with certain similar aspects. Just as species are united into genera, so are genera combined into families. The family name is not part of the scientific name, but is used to group those genera that have similar characteristics. This name can be readily recognized because, with few exceptions, it ends in *aceae*.

Following the scientific name is the name, or abbreviation of the name, of the person who first made that particular combination of genus and species, thus becoming the "authority" for that scientific designation. Occasionally a name in parentheses will be inserted to indicate that the authority in parentheses had used the specific epithet in combination with a different generic name for that plant; the authority following the parentheses made the present combination.

One or two Mexican common names, if any are known, accompany each entry in this handbook, for often a local common name will facilitate communication about a plant. Widely distributed and particularly useful

INTRODUCTION

plants may have several common names; unfortunately, the same common name may be applied to different plants. Use of the botanical name, of course, avoids confusion.

All entries are arranged alphabetically by family, by genus within family, and by species within genus. Each entry begins with a plant's scientific name, and is followed by its local common name or names, if known. Each plant described is indexed by both common and scientific name; family names also are indexed. Each plant or closely related group of plants is illustrated.

Plant Identification Key

The Plant Identification Key is designed to help the reader pick a particular plant from among the more than two hundred selected Mexican species described in this guide. The outstanding characteristics that attract attention to the plant have been used as a basis for the key; botanical terms are defined in the Glossary on page 361.

As do most botanical keys, this one comprises a series of contrasting couplets with a name of number at the end of each half couplet. The number indicates the next couplet to be read. Continuing in this way until a name is substituted for a number will result in identification of species or family, enabling the reader easily to locate the appropriate entry in the text. If the plant in question does not resemble the description or picture, either the plant is not included or a misinterpretation of some character led to selection of the wrong couplet. The numbers in parentheses in the left-hand column indicate the last couplet previously considered so that one may backtrack, if necessary.

THE KEY

1	Leaves needles, plants conifers	2
1	Leaves broader, or absent, not needles	3
2(1)	Needles in bundles	Pinaceae
2	Needles single	Taxodiaceae
3(1)	Plants palmlike	4
3	Plants not palmlike	5
4(3)	Leaves bipinnately compound, young leaves curled	Cyatheaceae
4	Leaves palmately lobed or pinnately compound	Palmae
5(3)	Plants succulent, spiny, cactuslike	Cactaceae
5	Plants not cactuslike	6
6(5)	Inflorescence arising from rosette of linear or lanceolate, usually fleshy leaves	Agavaceae
6	Rosette of linear leaves not formed	7

PLANT IDENTIFICATION KEY

7(6)	Plants leafless or flowering when leafless	8
7	Plants with conspicuous leaves	26
8(7)	Plants leafless or leaves reduced to small scales	9
8	Plants flowering when leafless	12
9(8)	Plants terrestrial, erect; stems cylindrical; latex present; desert habitat	10
9	Plants epiphytic; stems flat or cylindrical; no latex; tropical forest habitat	11
10(9)	Stems gray, pencil size; flowers in small cups .... Euphorbiaceae *Euphorbia*	
10	Stems gray, 1–1.5 cm in diameter, flowers in red irregular cups .... Euphorbiaceae *Pedilanthus*	
11(9)	Stems cylindrical, green, pendant, much branched; flowers small .... Cactaceae *Rhipsalis*	
11	Stems flattened; flowers large .... Cactaceae *Epiphyllum*	
12(8)	Flowers yellow	13
12	Flowers white, pink, red-purple, or blue	17
13(12)	Flowers many, small, clustered into heads .... Leguminosae (Mimosoideae) *Acacia farnesiana*	
13	Flowers in various inflorescences, not heads; large	14
14(13)	Petals free	15
14	Petals united	16
15(14)	Stamens 10 .... Leguminosae (Caesalpinioideae) *Cercidium*	
15	Stamens numerous ...... Cochlospermaceae	
16(14)	Flowers in open terminal panicles .... Bignoniaceae *Roseodendron*	
16	Flowers in condensed terminal clusters .... Bignoniaceae *Tabebuia chrysantha*	
17(12)	Flowers white	18
17	Flowers pink, red, purple, blue	19
18(17)	Large trees; bark white .... Convolvulaceae *Ipomoea arborescens*	

18	Shrubs or small trees; bark green ... Apocynaceae *Plumeria*
19(17)	Flowers red or blue 20
19	Flowers pink or purple 21
20(19)	Flowers red; stems spiny .... Fouquieriaceae
20	Flowers blue; trees without spines .... Zygophyllaceae *Guaiacum*
21(19)	Flowers pink 22
21	Flowers purple 25
22(21)	Petals united; stamens 4 ...... Bignoniaceae *Tabebuia rosea*
22	Petals free 23
23(22)	Stems with scattered spines .... Bombacaceae *Ceiba*
23	Stems without spines 24
24(23)	Flowers irregular (petals of different sizes); stamens 10 ..... Leguminosae (Papilionoideae) *Gliricidia*
24	Flowers large, regular; stamens numerous .... Bombacaceae *Bombax*
25(21)	Flowers in large, open panicles ... Bignoniaceae *Jacaranda*
25	Flowers in condensed clusters .... Bignoniaceae *Tabebuia rosea*
26(7)	Leaves compound 27
26	Leaves simple 46
27(26)	Leaves palmately compound 28
27	Leaves pinnate or bipinnately compound 29
28(27)	Petals free; stamens more than 4; fruit elongate; seeds hairy .... Bombacaeae
28	Petals united; stamens 4; fruit elongate; seeds winged, not hairy .... Bignoniaceae (in part)
29(27)	Leaves pinnately compound 30
29	Leaves bipinnately compound or appearing so 43
30(29)	One leaflet modified to a tendril 31
30	No tendril 33
31(30)	Flowers red or orange, irregular; vine 32
31	Flowers purple, regular; vine .... Polemoniaceae *Cobaea*

PLANT IDENTIFICATION KEY

32(31)	Flowers red, slightly irregular; leaves of 2 leaflets and tendril .... Bignoniaceae *Distictis*
32	Flowers orange, strongly irregular; leaves of 2 leaflets and tendril .... Bignoniaceae *Pyrostegia*
33(30)	Leaves even pinnate 34
33	Leaves odd pinnate 37
34(33)	Flowers small in terminal panicles; fruit dehiscent capsule; seeds winged .... Meliaceae
34	Flowers single or racemose; fruit and seeds not as above..35
35(34)	Leaves opposite .... Zygophyllaceae
35	Leaves alternate 36
36(35)	Flowers somewhat irregular; upper petal innermost .... Leguminosae Caesalpinioideae
36	Flowers strongly irregular; upper petal outermost .... Leguminosae (Papilionoideae) *Olneya*
37(33)	Flowers yellow, showy; petals united 38
37	Flowers not yellow, often small, not showy 39
38(37)	Flowers funnelform; spines absent ... Bignoniaceae *Tecoma*
38	Flowers wheel-shaped; spiny .... Solanaceae *Solanum rostratum*
39(37)	Flowers strongly irregular, showy, pink, red, or purple; fruit a legume 40
39	Flowers regular, small, not showy 42
40(39)	Flowers red; leaflets 3, large; seeds red .... Leguminosae (Papilionoideae) *Erythrina*
40	Flowers pink or purple 41
41(40)	Flowers in panicles, pink to purple .... Leguminosae (Papilionoideae) *Andira*
41	Flowers in racemes, pink or purple ..... Leguminosae (Papilionoideae) *Gliricidia* or *Sophora*
42(39)	Bark peeling in paper-thin-layers; few-flowered inflorescences .... Burseraceae
42	Bark not peeling; flowers in large panicles .... Anacardiaceae *Pseudosmodingium* or *Schinus*

43(29)	Petals united; fruit a capsule	44
43	Petals not united; fruit a legume; seeds not winged	45
44(43)	Fruit a flattened capsule; seeds winged; flowers blue, blue-violet .... Bignoniaceae *Jacaranda*	
44	Fruit a capsule; seeds not winged; flowers red or orange; plant herbaceous .... Scrophulariaceae *Lamourouxia*	
45(44)	Flowers small in spikes or capitate; petals not showy; stamens conspicuous .... Leguminosae Mimosoideae	
45	Flowers in racemes or panicles; petals showy, irregular, upper petal innermost .... Leguminosae Caesalpinioideae	
46(26)	Leaves lobed	47
46	Leaves not conspicuously lobed	58
47(46)	Leaves simple, pinnately lobed	48
47	Leaves simple, palmately lobed	52
48(47)	Leaves irregularly pinnately lobed; plants herbaceous, spiny	49
48	Leaves regularly pinnately lobed; plants shrubs, trees, or vines	50
49(48)	Petals 6, free, pale to dark yellow; stamens numerous; leaves glaucous .... Papaveraceae *Argemone*	
49	Petals 5, united; stamens 5 .... Solanaceae *Solanum rostratum*	
50(48)	Large tropical vine, leaves large, perforate; flowers on heavy stalk surrounded by bract .... Araceae *Monstera*	
50	Shrub or tree	51
51(50)	Flowers small, paniculed; petals inconspicuous .... Papaveraceae *Bocconia*	
51	Flowers showy; petals 5, united .... Solanaceae *Solanum* (in part)	
52(47)	Leaves opposite, 2 diverging lobes; flowers yellow; fruit fuzzy balls .... Zygophyllaceae *Larrea*	
52	Leaves alternate, several lobes	53
53(52)	Flowers perfect, yellow, single, racemose, or as composite heads	54

PLANT IDENTIFICATION KEY

53	Flowers unisexual, with occasional perfect flower; leaves 5 to 7-lobed 55
54(53)	Flowers large, single, or racemose; fruit dark brown capsule; seeds hairy; leaves 5-lobed .... Cochlospermaceae
54	Flowers clustered into composite heads, ray flowers present; leaves 5 to 7-lobed .... Compositae *Senecio praecox*
55(53)	Leaves deeply many-lobed, conspicuously white below, green above; long-petioled .... Moraceae *Cecropia*
55	Leaves deeply or shallowly 5 to 7-lobed, not prominently colored 56
56(55)	Leaves 5 to 7-palmate lobed; lobes pinnate-lobed; fruit melonlike along stem .... Caricaceae
56	Leaves lobed; characters not as above 57
57(56)	Flowers perfect, clustered into heads, forming a panicle; fruit black berries .... Araliaceae
57	Flowers unisexual, male below female; fruit prickly, 2 or 3-seeded .... Euphorbiaceae *Ricinis*
58(46)	Leaves conspicuously parallel-veined 59
58	Leaves net-veined 67
59(58)	Plants herbaceous; flower parts in 3's or 6's 60
59	Plants woody shrubs or trees; flower parts in 4's or 5's 66
60(59)	Vine; flowers umbellate, pendant; fruit capsule with red seeds .... Amaryllidaceae *Bomarea*
60	Plant not a vine 61
61(60)	Leaves arising from a subterranean bulb or aerial pseudobulb, no petiole 62
61	Leaves petioled; no bulb 65
62(61)	Flowers white, or white and green 63
62	Flowers red or red-orange; stamens 3 or 6 64
63(62)	Flowers star-shaped; stamens 6 .... Amaryllidaceae *Milla*
63	Flowers very irregular; stamens modified .... Orchidaceae *Epidendrum*

16

PLANT IDENTIFICATION KEY

64(61)	Flowers red at right angles to stem; stamens 6 .... Amaryllidaceae *Sprekelia*
64	Flowers orange-red, erect; stamens 3 .... Iridaceae
65(61)	Stem tall, bananalike; flowers inconspicuous, subtended by showy bracts .... Musaceae
65	Stem short, petiole long, blade small; bract green, flower showy .... Strelitziaceae
66(59)	Leaves with 5 veins from base to apex connected by parallel veins .... Melastomataceae
66	Leaves with midrib, parallel veins extending from midrib to margin .... Apocynaceae
67(58)	Plants distinct vines 68
67	Herbs, shrubs, or trees, occasionally scandent 78
68(67)	Flowers minute, inconspicuous, clustered into spikes .... Piperaceae
68	Flowers larger, conspicuous; inflorescence varies 69
69(68)	Showy bracts present 70
69	Showy bracts absent 71
70(69)	Flowers smaller than subtending showy bract, clusters of 3 .... Nyctaginaceae *Bougainvillea*
70	Flowers exceeding bracts; inflorescence raceme .... Convolvulaceae *Exogonium*
71(69)	Leaves opposite 72
71	Leaves alternate 75
72(71)	Petals united 73
72	Petals free 74
73(72)	Petals pink or white; fruit paired, latex present .... Asclepiadaceae *Cryptostegia*
73	Petals and sepals blue; fruit inconspicuous; latex absent .... Verbenaceae *Petrea*
74(72)	Petals clawed, yellow; inflorescence raceme; fruit 4-winged .... Malphigiaceae *Mascagnia*

PLANT IDENTIFICATION KEY

74	Petals small, red; stamens showy, red or greenish yellow; inflorescence one-sided; fruit 4-winged .... Combretaceae *Combretum*
75(71)	Perianth parts free .... Polygonaceae *Antigonon*
75	Perianth parts united 76
76(75)	Flowers regular, funnelform; seashore plants, often rooting at nodes; leaves round .... Convolvulaceae *Ipomoea pes-caprae*
76	Flowers irregular 77
77(76)	Flowers extremely large, very irregular; resembling oversized smoking pipe .... Aristolochiaceae
77	Flowers tubular with irregular-sized lobes; climbs by twining petioles and pedicels .... Scrophulariaceae *Maurandya*
78(67)	Flowers unisexual; fruit acorn .... Fagaceae *Quercus*
78	Flowers perfect or unisexual; fruit not acorn 79
79(78)	Plants floating aquatics .... Pontederiaceae *Eichhornia*
79	Plants not aquatics 80
80(79)	Individual flowers clustered into heads, balls, or figlike fruits 81
80	Individual flowers not distinctly clustered into heads, balls, or figlike fruits; inflorescences various 85
81(80)	Fruits small figs .... Moraceae *Ficus*
81	Fruits not figlike 82
82(81)	Flowers clustered into a flattened head, surrounded by bracts; ovary inferior .... Compositae
82	Flowers clustered into round balls 83
83(82)	Flowers unisexual; balls axillary; fruit fleshy, one seed; tree of warm moist tropics .... Moraceae *Brosimum*
83	Flowers perfect; balls in axillary or terminal racemes or panicles; seeds many 84
84(83)	Inflorescence axillary or terminal raceme; plants of coastal marshes .... Combretaceae *Conocarpus*
84	Inflorescence terminal panicle; plants not of marshes .... Loganiaceae

PLANT IDENTIFICATION KEY

85(80)	Leaves opposite	86
85	Leaves alternate	100
86(85)	Perianth parts free, occasionally at end of floral tube	87
86	Perianth parts united	92
87(86)	Petals yellow, clawed; inflorescence a raceme or panicle; fruit a dry capsule or fleshy	Malpighiaceae
87	Petals white or red, prominent, inconspicuous, or absent; sepals green, red or yellowish; flowers 1, 2, or many per peduncle	88
88(87)	Sepals green or yellow; petals white; seed 1; flowers 2 to many per peduncle; plants of coastal marshes	89
88	Sepals red; not of coastal marshes	90
89(88)	Sepals yellow; petals white; 2 flowers per peduncle; ovary superior	Rhizophoraceae
89	Sepals green; petals white; inflorescence a panicle; ovary inferior	Combretaceae *Laguncularia*
90(88)	Petals absent; sepals red, petal-like; several flowers per peduncle; plants parasitic	Loranthaceae
90	Petals present, red or white; floral tube present	91
91(90)	Floral tube spurred; petals white; fruit a capsule, many seeds; ovary superior	Lythraceae
91	Floral tube straight; petals red; fruit a berry; ovary inferior	Onagraceae
92(86)	Flowers regular	93
92	Flowers irregular, sometimes only slightly	96
93(92)	Ovary inferior; flower parts 4's or 5's	Rubiaceae
93	Ovary superior	94
94(93)	Flowers unisexual; male flowers in dense clusters, stamens prominent; female flowers paniculate; fruit 5 angles, glandular; curved spines	Nyctaginaceae *Pisonia*
94	Flowers perfect; herbs, shrubs, or trees without spines	95
95(94)	Flowers yellow, funnelform; fruit paired; latex present	Apocynaceae *Stemmadenia*

19

PLANT IDENTIFICATION KEY

95	Flowers red, petals reflexed; fruit follicle; seed hairy; latex present .... Asclepiadaceae *Asclepias*	
96(92)	Plants herbaceous, sometimes semi-shrubby	97
96	Shrubs or trees	98
97(96)	Flowers on long pedicels; fruit a many-seeded capsule; stamens 4 .... Gesneriaceae	
97	Flowers on short pedicels, clustered into groups; fruit 4 one-seeded nutlets; stamens 2 .... Labiatae	
98(96)	Flowers strongly irregular; corolla bilabiate .... Acanthaceae *Justicia*	
98	Corolla not strongly irregular	99
99(98)	Fruit a capsule, many seeded; flowers large, tubular, yellow .... Acanthaceae *Ruellia*	
99	Fruit a berry or 1-seeded capsule .... Verbenaceae	
100(85)	Petals absent	101
100	Petals present, sometimes small	105
101(100)	Stamens and sepals red; stamens fused into handlike structure; large tropical tree .... Sterculiaceae *Chiranthodendron*	
101	Stamens not as above	102
102(101)	Flowers perfect or unisexual, small, yellow green	103
102	Flowers unisexual	104
103(102)	Spiny desert shrub; fruit fleshy berry .... Ulmaceae *Celtis*	
103	Desert shrub; no spines; fruit winged .... Sapindaceae *Dodonea*	
104(102)	Fruit in grapelike clusters; flowers small, white .... Polygonaceae *Coccoloba*	
104	Fruit various, not as above; latex present ... Euphorbiaceae	
105(100)	Petals united	106
105	Petals free	118
106(105)	Flowers irregular	107
106	Flowers regular	110

107(106) Leaves linear .... Bignoniaceae
107 Leaves broad 108
108(107) Herbaceous 109
108 Gray shrub; flowers purple
 .... Scrophulariaceae *Leucophyllum*
109(108) Corolla narrow, strongly irregular; anthers fused; fruit
 a capsule .... Lobeliaceae
109 Corolla broad tubular; fruit with 1 or 2 curved
 beaks .... Martyniaceae
110(106) Inflorescence scorpioid 111
110 Inflorescence various 112
111(110) Flowers white, deeply funnelform ..... Boraginaceae *Cordia*
111 Flowers purple, shallowly
 funnelform .... Hydrophyllaceae *Wigandia*
112(110) Flowers small, numerous in axillary clusters; latex
 present .... Sapotaceae
112 Flowers few in axillary clusters 113
113(112) Corolla shallow funnelform or
 discoid .... Solanaceae *Solanum*
113 Corolla tubular, urn-shaped, or deep
 funnelform 114
114(113) Corolla urn-shaped .... Ericaceae
114 Corolla tubular or funnelform 115
115(114) Corolla funnelform, white 116
115 Corolla tubular 117
116(115) Large tree, flowers not
 pendant .... Convolvulaceae *Ipomoea arborescens*
116 Shrub or small tree, flowers
 pendant .... Solanaceae *Datura*
117(115) Corolla yellow .... Solanaceae *Nicotiana*
117 Corolla red .... Polemoniaceae *Loeselia*
118(105) Stems and branches spiny 119
118 No spines 120

PLANT IDENTIFICATION KEY

119(118) Spines abundant, clustered; flowers
   white .... Cactaceae *Pereskia*
119 Spines 1 per node; flowers red or
   cream .... Fouquieriaceae
120(118) Flowers with more than 10 stamens 121
120 Stamens 10 or fewer 124
121(120) Flowers single, white or yellow 122
121 Flowers in terminal clusters white or pink 123
122(121) Flowers large, white; stamens many, free .... Magnoliaceae
122 Flowers large, yellow; stamens many, fused into a column
   around style .... Malvaceae *Bakeridesia*
123(121) Inflorescence a raceme or panicle; red-brown fruiting capsule
   with long soft prickles .... Bixaceae *Bixa*
123 Inflorescence dense corymb; fruit a woody capsule; no
   prickles .... Rosaceae *Vauquelinia*
124(120) Small unbranched rubbery shrub; flowers unisexual; male
   flowers clustered; female single
   .... Euphorbiaceae *Jatropha*
124 Branched shrub or tree; flowers small 125
125(124) Inflorescence large panicle; large tree; fruit edible
   mango .... Anacardiaceae *Mangifera*
125 Inflorescence few-flowered 126
126(125) Inflorescence few-flowered axillary or terminal panicle;
   fruit globose with numerous prickles .... Tiliaceae
126 Inflorescence 1- or 2-flowered; fruit not prickly 127
127(126) Flowers red; leaves sharp-pointed; fruit woody
   balls .... Theophrastaceae *Jacquinia*
127 Flowers white or cream; fruit warty or
   smooth .... Sterculiaceae

The Species

Acanthaceae—Acanthus Family

Justicia spicigera Schl.
Hierba Azul or Mohintli

A widespread native of Mexico, *Justicia spicigera* is a popular cultivated plant both there and in the United States. Its sprightly red or orange tubular flowers are borne on axillary or terminal, usually one-sided racemes. As only a few flowers on the raceme open at the same time, the blooming period is extended. The narrow, straight, or slightly arched tubular corolla is two-lobed, the lower one recurved and coiled. Two stamens are tucked under the tip of the upper lobe, from which the slender pistil extends. After pollination the corolla separates from the short calyx, leaving the long style that persists as the elongated fruit develops. Viscid, lance-oblong leaves and stems are covered with short soft hairs. These leaves, when boiled, produce a blue solution used to whiten clothes. The Nahuatl name "mohintli" translates to "blue," referring to this property, as does the Mexican common name "hierba azul" (blue plant).

Widely cultivated and a favorite flower of hummingbirds, *Justicia* grows natively from Durango and San Luis Potosí south to Veracruz, Oaxaca, and Chiapas. References to this plant also can be found under the name *Jacobinia*.

Ruellia speciosa (Nees) Lindau

As the majority of *Ruellia* tend to be shades of pink, purple, or blue, *R. speciosa*, with its conspicuous pale yellow funnelform flowers, is an unusual, showy, small shrub. One to three large flowers with long,

ACANTHACEAE—ACANTHUS FAMILY

Justicia spicigera

SHRUB 1–1.5 m (3–5 ft) tall. LEAVES 6–17 cm (2.5–6.5 in) long, lance-oblong to ovate, short petiolate, pubescent. FLOWER 3–3.5 cm (1.25–1.5 in) long, red or orange, 2-lipped; stamens 2. FRUIT a capsule.

narrow tubes arise from the axil of a leaf. The throat of the tube markedly inflates after extending beyond the deeply cut, long-attenuated calyx lobes. Five rounded, recurving corolla lobes spread to reveal four large stamens nestled in the throat opening. Leaves on long petioles are dark green above and lighter green below, with wavy margins and short hairs on both sides. They are viscid to the touch. The dark brown capsular fruit explodes at maturity, dispersing its seeds in many directions.

Many species of *Ruellia* grow in Mexico, but this one favors the states of Hidalgo, México, Puebla, and Oaxaca.

ACANTHACEAE—ACANTHUS FAMILY

Ruellia speciosa
SHRUB 1−2 m (3−6.5 ft) tall. LEAVES 4−7.5 cm (1.5−3 in) ovate or oblong-ovate, sticky short hairs; long petioles. FLOWER 6 cm (2.5 in) long, pale yellow, inflated funnel-form; calyx 2.5−3 cm (1−1.25 in) long. FRUIT 1−1.5 cm (.5−.75 in) long, explosive capsule.

Agavaceae—Agave Family

Agave Species

Agave, the first genus of the Agave Family considered here, originally was placed by botanists in the Amaryllis Family (Amaryllidaceae). Agaves are short-stemmed plants with many usually elongated leaves in a rosette formation. Some forms produce rhizomes, resulting in the production of small plantlets (suckers) under the canopy of the leaves of the parent. Frequently applied to some of the cultivated species is the common name century plant, alluding to the long time required for production of a flowering stalk; a period of eight to twenty years, depending on the species, is necessary for the plant to accumulate sufficient food reserves to induce the reproductive cycle. In most *Agave* species, the plant dies after it has flowered.

The spirally arranged leaves are thick, succulent water storage organs, as well as protective weapons. They have a thick epidermis and cuticle to prevent water loss and are channeled so that even slight rains are directed to the root area. Each leaf is tipped with a stout terminal spine and most frequently with hard, sharp, lateral teeth; the teeth, when present, are variable in the extreme. It is this armament that protects the tender flowering stalk from destruction by animals and renders the larger species suitable for fences if planted closely enough to form impenetrable barriers.

Inflorescences assume two general forms, although intermediates occur. Spicate racemes produce few to many sessile or nearly sessile flowers along a tall central axis. In paniculate inflorescences, large compound umbels of erect flowers are produced terminally on lateral branches extending at right angles from the central axis. In both types there is a transition from leaves of the rosette to modified leaves, or bracts, along the flowering axis to bractlets subtending the various branches and ultimate floral pedicels.

All *Agave* flowers are perfect, with a three-celled inferior ovary. The perianth contains six similar or dissimilar segments attached to the top of the long or short floral tube. Six stamens variously attached to the floral tube or to the base of the perianth segments are prominently exserted; each is tipped with a large versatile anther. From the central base of the floral tube—essentially the top of the ovary—arises a single heavy style.

Bats in search of pollen or nectar, which may be abundant in the paniculate, erect flowers, are a primary, but not the only, pollinating agent. Fruits resulting from the pollination are three-celled, woody oblong or ovoid capsules with many black discoid seeds.

Agaves have played an important role in the human economy from ancient times, as a source of food, drink, and fiber. The meristematic tissue of the short stem and leaf bases contains abundant starch, or sugar. Many are edible, but care must be taken to avoid those species

AGAVACEAE — AGAVE FAMILY

that produce sapogenins. The tender flowering shoots of most species are edible, as are the flowers. Agave leaves also are cut and trimmed as food for cattle.

Drinks, both sweet and alcoholic, are a product of the agave. *Aguamiel* is a fresh, unfermented juice expressed from wild or cultivated plants; the alcoholic beverage *pulque* is produced by fermentation, and *mescal* by distillation.

Cordage processed from the long, strong fibers of agave leaves is used in baskets, mats, nets, bags, blankets, and anything else in which a rough, heavy fiber can be used.

Much is yet to be learned about agave chemistry, including why some people develop such a severe rash from contact with the juice, or what on the spines causes such a serious and prolonged injury when one is impaled.

It is possible in a work of this type to cover only a few of the most widely distributed or economically important species; of the approximately 120 species of *Agave* recorded for Mexico, eight are discussed. The comprehensive publication on the agaves by Howard Scott Gentry, *Agaves of Continental North America*, will be of interest to those desiring more detailed information on all aspects of this important plant.

Agave americana L.
Maguey or Century Plant

Agave americana, the century plant so frequently seen in stylized drawings of Mexico, is a large rosette of light gray-green, glaucous leaves, lanceolate in shape, bearing many sharp marginal teeth and terminating in a coarse brown spine. Often the leaves bear the marginal imprint of an overlapping leaf acquired while still a part of the tightly rolled, conical central bud. The flowering stalk will extend to its full development in two to four months. The straight stalk has many small

Agave americana
ROSETTE 1−2 m (3−6.5 ft) tall, 2−3.7 m (6.5−13 ft) wide. LEAVES 1−2 m ▶ (3−6.5 ft) long, 15−25 cm (6−10 in) wide, teeth 5−10 mm long, spine 3−5 cm (1.25−2 in) long. FLOWER STALK 5−9 m (16−30 ft) tall, 15−35 paniculate umbels. FLOWER 7−10 cm (2.75−4 in) long.

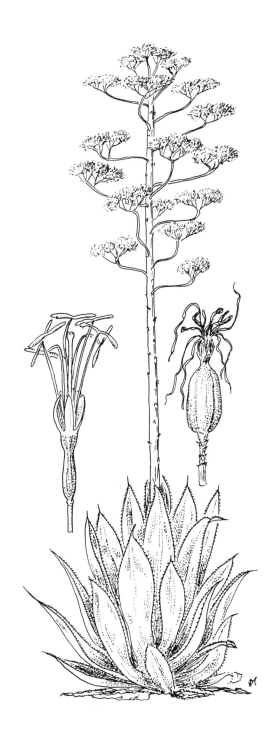

29

AGAVACEAE — AGAVE FAMILY

bracts and forms 15 to 35 horizontal branches. Numerous erect yellow flowers in compound umbels occupy the ends of the branches. Seeds are produced in woody capsules, but reproduction also is accomplished by the formation of numerous suckers.

This species, worldwide in distribution, has been in cultivation for such a long time that its original home is unknown. In addition to being widely distributed as a cultivated ornamental, it is found commonly in the eastern Sierra Madre from Coahuila to Tamaulipas and south to Oaxaca. Forms also grow in Baja California, Sonora and Jalisco.

Agave angustifolia Haw.
Mescal de Maguey

As both a native and cultivated plant, *Agave angustifolia* is widely distributed in Mexico. It forms a short-stemmed rosette of long, light green to glaucous gray, rigid leaves with small reddish-brown to dark brown marginal teeth and a dark brown terminal spine. Clusters of green to yellow flowers terminate the 10 to 20 horizontal branches of the open panicle. The fertile flowers produce broadly ovoid, dark brown, woody capsules and occasionally a few bulbils. Large colonies of *A. angustifolia* form as a result of the development and growth of rhizomatous suckers.

The short stem of *A. angustifolia* stores a great amount of starch, making this species important in the production of mescal; in the region of Oaxaca a glaucous, blue-gray variety is cultivated especially for this purpose. It resembles "mescal azul," the *A. tequilana* of the Jalisco area, but differs in its spine and teeth forms. In the state of Sonora the wild specimens of *A. angustifolia* are gathered to produce "mescal bacanora," a regionally famous mescal that takes its name from the area where it is made.

Widely ranging *A. angustifolia* in all its varieties can be found from Sonora on the west and Tamaulipas on the east, southward throughout Mexico to Costa Rica.

Agave angustifolia
ROSETTE 20–60 cm (7–24 in) tall. LEAVES 60–120 cm (24–47 in) long, 3.5–10 ▶ cm (1.5–4 in) wide; armed. FLOWER STALK 3–6 m (10–20 ft.) tall, open panicle. FLOWER 5–6.4 cm (2–2.75 in) long. CAPSULE 5 cm (2 in) long, 3 cm (1.25 in) wide.

AGAVACEAE — AGAVE FAMILY

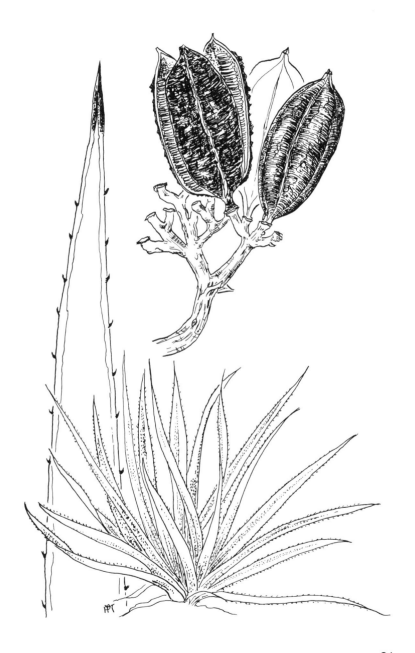

31

AGAVACEAE — AGAVE FAMILY

Agave fourcroydes Lem.
Sacqui or Gray Agave

ROSETTE 1–2 m (3–6.5 ft.) tall. LEAVES 1.5–2.5 m (5–8 ft) long, 8–10 cm (3–4 in) wide; teeth 3–6 mm, spine 2–3 cm (.75–1.25 in) long. FLOWER STALK 5–6 m (16–20 ft) long, 10–18 paniculate umbels. FLOWER 60–70 mm long; perianth 16–18 mm long.

Economically one of Mexico's important plants, *Agave fourcroydes* has been cultivated so long as an ornamental and fiber-producing plant that its original home is unknown. It is recognized by its vigorous growth and by its long, narrow, rigid, light green leaves. A dark brown terminal spine and small, evenly spaced, lateral, spinelike teeth arm the leaves. Long panicles of greenish-yellow sterile flowers are produced; seed capsules do not form. In their stead, bulbils (small plantlets) develop in axils of the floral bracts after flowering. They persist for a few years, after which they fall to the ground to develop new plants if soil covers them. The slow growth of the bulbils renders the use of rhizomatous suckers a more expedient method of obtaining propagating stock.

Plantations of *A. fourcroydes* will be seen in the warm areas of southern Tamaulipas, Veracruz, and Yucatán. The fiber derived from the leaves is used for such coarse material as binding twine and ship ropes. In recent years cortisone has been synthesized from the sapogenin. which is part of the waste material of fiber extraction.

Agave lechuguilla Torr.
Lechuguilla

Readily recognized by its common name "lechuguilla," which also has been applied as the specific epithet, this species is one of the most widely distributed and most abundant of the agaves. The small rosettes are widely suckering, as well as freely seeding, and produce but few linear-lanceolate, light green or yellow-green leaves. These are armed with a gray terminal spine and small, evenly spaced, brown or light gray, downward-slanting teeth. On dry mature leaves this spiny margin is easily removed. Spicate flower stalks have short-pedicelled, yellow or reddish-tinged flowers in clusters of twos or threes.

Agave lechuguilla, also found in Texas and New Mexico, ranges

AGAVACEAE — AGAVE FAMILY

Agave lechuguilla

ROSETTE small. LEAVES 25–50 cm (10–20 in) long, 2.5–4 cm (1–1.5 in) wide; teeth 2–5 mm, spine 1.5–4 cm (.75–1.5 in) long. FLOWER STALK 2.5–3.5 m (8–12 ft) tall, spicate. FLOWER 3–4.5 cm (1.25–2 in) long. CAPSULE 18–25 mm long, 11–18 mm wide.

from the northern border states of Mexico south to Hidalgo and the state of Mexico. It is partial to limestone-derived soils and absent in volcanic areas. This species is the source of the hard fiber "istle" used for rope, twine, brushes, and other articles. Because it does not respond to cultivation, leaves are harvested by hand from wild plants on the rocky hillsides. Fibers also are extracted by the use of hand tools. Saponin is present in *A. lechuguilla*, rendering it poisonous to cattle.

Agave mapisaga Trel.
Maguey Mapisaga

ROSETTE 2–2.5 m (6.5–8 ft) tall. LEAVES 185– 250 cm (72–100 in) long, 19–25 cm (7.5–10 in) wide; teeth 2–5 mm long, spine 3–5 cm (1.25–2 in) long. FLOWER STALK 7–8 m (23–26 ft) or more tall; 20–25 paniculate umbels. FLOWER 8–10 cm (3–4 in) long. CAPSULE 6–6.5 cm (2.5–2.75 in) long, 2 cm (.75 in) wide.

Agave mapisaga, another of the *pulque* plants, forms massive open rosettes of long, narrow, pale green or pale gray glaucous leaves armed with small brown marginal teeth and a dark or gray-brown spine. The large flowering stalk has the usual succulent, closely set bracts and produces in the upper portion 20 to 25 branches, each ending in a heavy, compound, umbellate panicle. Large flowers, reddish when in bud, open with yellow perianth parts over the green ovary.

This agave is found in the same area of the Central Plateau as is *A. salmiana*, which also is used in *pulque* production. The two species frequently are cultivated together, but *A. mapisaga* is always in the minority, possibly because it produces fewer suckers. It is distinguished from *A. salmiana* by its longer, linear leaves, which lack a sigmoid curve.

Agave salmiana Otto ex Salm.
Maguey de Pulque

A very important *pulque*-producing species, *Agave salmiana* is found both in extensively cultivated and wild populations. It has massive closely suckering rosettes with short thick stems. Broadly lanceolate,

AGAVACEAE — AGAVE FAMILY

Agave salmiana
ROSETTE 1.5–2 m (5–6.5 ft) tall. LEAVES 1–2 m (3–6.5 ft) long, 20–35 cm (7.5–13.5 in) wide; armed, teeth 5–10 mm long, spine 5–10 cm (2–4 in) long. FLOWER STALK stout, 7–8 m (23–26 ft) tall, 15–20 paniculate umbels. FLOWER 8–11 cm (3–4 in) long. CAPSULE 5.5–7 cm (2.25–2.75 in) long, 2 cm (.75 in) wide.

AGAVACEAE — AGAVE FAMILY

green or glaucous gray leaves with sigmoidly bent, acuminate tips are well-armed. The long, stout, dark brown spine often extends to midblade as a horny margin. Large brown or gray-brown teeth mark the slightly undulating edge of the leaf. The stout flowering stalk has thick fleshy bracts closely overlapped, and 15 to 20 branches. Terminal compound umbels are made up of large flowers, each with a thick, fleshy, yellow perianth surmounting a green ovary.

Agave salmiana is abundant throughout the Central Plateau region of Mexico from Michoacán, Guanajuato, Queretaro, and San Luis Potosí south to Puebla and Morelos. It has been estimated that 75 percent of the *pulque* consumed in this area is produced by this species. *A. salmiana* is related to *A. mapisaga*, with which it frequently is cultivated. It can be distinguished from that species by its broad, heavy, well-armed leaves with narrowing sigmoid tips.

Agave sisalana Perrine
Sisal

Agave sisalana, one of the fiber-producing agaves, is recognized by its long green leaves, which are unarmed or edged with very small teeth, but tipped with a short, dark brown, conical spine. In this species, as in *A. fourcroydes*, the basal leaves are harvested for their fibers, resulting in the formation of an elongated, trunklike stem. If a flowering stalk develops, it will have 10 to 15 lateral branches bearing umbellate clusters of greenish-yellow flowers. Although looking perfectly normal, the flowers are sterile and produce small bulbils in place of fruiting capsules or seeds. Rhizomatous suckers serve as the primary means of reproduction because the bulbils develop extremely slowly.

The origin of *A. sisalana* is unknown; it may represent a hybrid between two species. It has long been thought to have originated in the Yucatán, but no wild plants have been found in that area. It is now worldwide in distribution as a cultivated, tropical, fiber plant, and is found in plantations and fence rows in Chiapas. Cultivation of this species is expanding in Yucatán to replace the armed leaf of *A. fourcroydes*.

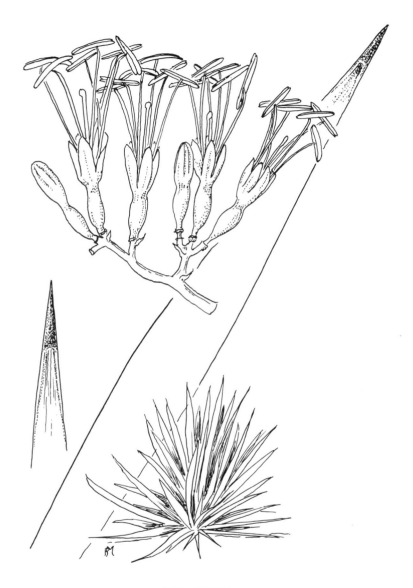

Agave sisalana

ROSETTE stem 1.2−2 m (3.5−6.5 ft) tall. LEAVES 90−130 cm (35−50 in) long, 9−12 cm (3.5−5 in) wide; teeth lacking, spine 2−2.5 cm (.75−1 in) long. FLOWER STALK 5−6 m (16−20 ft) tall, 10−15 paniculate umbels. FLOWER 5.5−6.5 cm (2.25−2.75 in) long, sterile.

AGAVACEAE — AGAVE FAMILY

Agave tequilana Weber
Mescal Azul or Maguey Tequilero

One of Mexico's most famous magueys, *Agave tequilana* is the source of tequila, the mainstay of the Margarita cocktail. The short thick stems store abundant quantities of starch. Lanceolate, blue-gray to blue-green leaves are armed with evenly sized and regularly spaced light to dark brown teeth and tipped with a short, conical, dark brown spine. The tall panicle has 20 to 25 large branches, bearing dense, paniculate umbels of green flowers with roseate stamens. The narrow perianth parts turn brown and wither quickly after anthesis. Reproduction is by seed or by abundant suckers.

This species is closely related to *A. angustifolia*, from which it differs by its larger leaves, thicker stems, and heavier panicle of larger flowers. All these differences are a matter of degree, making separation of these two species difficult.

Agave tequilana is primarily a cultivated form, possibly derived from wild populations near the city of Tequila in Jalisco, about 40 miles northwest of Guadalajara where thousands of acres of *A. tequilana* are planted. At the age of approximately eight years, the tequila plant is harvested by removing its leaves; the remaining stem, called the *cabeza*, which weighs from 60 to 120 pounds, is taken to the factory to be placed in a steam cooker to convert the stored starch to sugar. Maceration and pressing operations extract the sweet juice, which is allowed to ferment; distillation produces the raw tequila. Many of the tequila factories welcome visitors and conduct regular tours.

Beaucarnea gracilis Lem.
Sotolin

Its enormously swollen, fissured base makes for easy identification of this tree, which somewhat resembles a yucca. Stout naked branches arising from the bulbous basal stem are topped with a tuft of stiff, spiky, linear leaves. From this radiating burst of leaves dense panicles of unisexual flowers develop. The flower consists of six whitish perianth parts, the male flower containing six stamens and the female a three-

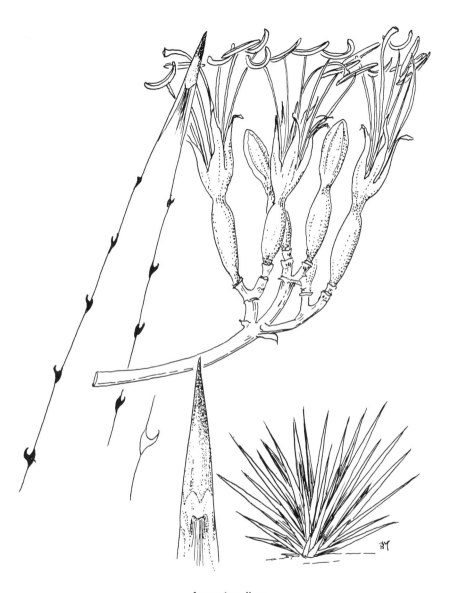

Agave tequilana

ROSETTE stems short 30–60 cm (12–24 in) long. LEAVES 90–120 cm (35–48 in) long, 8–12 cm (3–5 in) wide, armed, teeth 3–6 mm long, spine 1–2 cm (.5–.75 in) long. FLOWER STALK 5–6 m (16–20 ft) tall; 20–25 large, paniculate umbels. FLOWER 68–75 mm long.

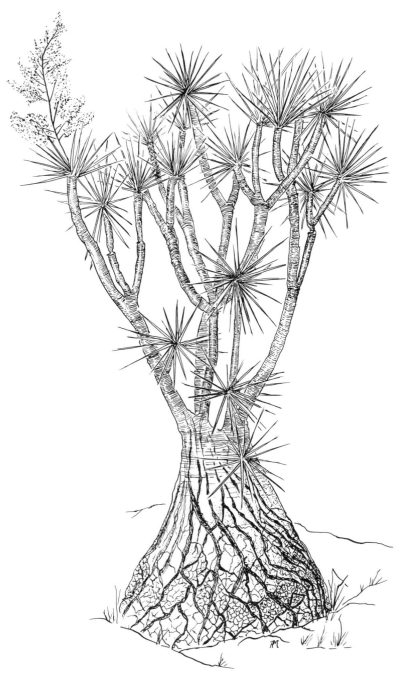

AGAVACEAE — AGAVE FAMILY

lobed pistil. Small fruits develop a wing from each of the angles of the ovary.

This species of *Beaucarnea* is restricted to the state of Puebla. Other species with similarly limited distributions grow in San Luis Potosí, Veracruz, Yucatán, and Oaxaca. All resemble one another, variations occurring in fruit or leaf size.

YUCCA SPECIES

Dense panicles of large white or creamy white flowers arising from the center of a cluster of long, narrow leaves indicate *Yucca*, a characteristic floral member of drier areas. Plants may develop into large tree forms, or may be straight unbranched columns, each with a leaf cluster terminating the stems. Narrow, swordlike leaves are flat or slightly concave, with sharply pointed tips and smooth, finely toothed or fibrous margins. Despite variations in plant form, all yucca flowers have six fleshy perianth parts, six stamens, and a superior ovary with three cells; each panicle has the potential of producing many large fruits; these may be indehiscent, fleshy berries, or hard, dry, dehiscent capsules. Numerous black, flat seeds usually are formed in each of the three cells.

Yuccas exhibit an interesting symbiotic relationship between a plant and an insect. The larva of a moth feeds on the developing yucca seeds; to ensure that seeds develop, the female moth gathers the sticky pollen of the yucca into balls and carries it to the stigmatic area of another flower. After packing the ball of pollen into the style cavity, the female moth deposits her eggs in the young ovary. Upon hatching, the larva feeds on the developing ovules, but eats only a few of the many potential seeds. Eventually the mature larva burrows out of the still tender, succulent fruit and drops to the ground to complete its life cycle. The yucca fruit continues its development with maturation of the many remaining seeds. Each mature yucca fruit displays one or more scars as a result

Beaucarnea gracilis

◀ TREE 6–12 m (20–40 ft) tall, base swollen 2–7 m (6.5–23 ft) circumference. LEAVES 25–40 cm (10–16 in) long, 4–7 mm wide. FLOWER small in dense panicles, dioecious, white. FRUIT 3–5 mm long, winged.

AGAVACEAE—AGAVE FAMILY

of the exiting larvae. Without the moth, the yucca would not be pollinated; without the pollinated flower, the moth larva would have nothing to eat.

Yuccas sometimes have been categorized as members of the Lily Family (Liliaceae), but more recent studies place yuccas in the Agave Family.

UNBRANCHED YUCCAS

Among the columnar, unbranched, or few-branched forms of yucca, there are four species of wide distribution in Mexico: *Y. carnerosana*, *Y. rigida*, *Y. torreyi*, and *Y. treculeana*.

Yucca carnerosana (Trel.) McKelvey
Palma Samandoca

Yucca carnerosana, a tall, usually unbranched species, is topped with a symmetrical, hemispherical tuft of long rigid leaves; heavy fibers develop along leaf margins. The stem of the plant is protected by a skirt of dead leaves. The long thick inflorescence stalk elevates the dense ellipsoid panicle of showy, pleasantly fragrant flowers above the cluster of leaves. The flowering period is from March to April, with an occasional off-season display. The fleshy fruits are indehiscent. A plant of northern Mexico, this yucca grows in Chihuahua, Coahuila, Nuevo Leon, Tamaulipas, San Luis Potosí, and Zacatecas, with dense stands in northern San Luis Potosí.

* * *

Yucca rigida (Engelm.) Trel. has short (30–60 cm [11.75–23 in] long), finely denticulate leaves on its columnar stem. The stalk supporting the dense, many-flowered panicle is so short and completely enclosed by the leaves that only about half of the 4 to 6 cm-long, creamy white flowers show above the leaves during the flowering period of March and April. Small, capsular fruits 5 cm-long split open at maturity along three sutures for dispersal of the black discoid seeds. These woody fruits are persistent on the plants for some time after the seeds have

AGAVACEAE — AGAVE FAMILY

Yucca carnerosana

UNBRANCHED column 1.5–6 m (5–20 ft) tall. LEAVES .5–1 m (1.5–3 ft) long, 5–7.5 cm (2–3 in) wide. FLOWER 6–10 cm (2.5–4 in) long; inflorescence large, extending well above leaves. FRUIT 5–7.5 cm (2–3 in) long, 4 cm (1.5 in) in diameter, fleshy, indehiscent.

been distributed. This species grows in Durango, Chihuahua, and Coahuila.

* * *

Yucca torreyi Shafer is, for the most part, an unbranched, columnar type that may form clusters of stems. Long leaves with heavy filamentous margins tend to engulf half or more of the panicle on its short stalk. The relatively large (8 cm [3.25 in] long), cream-colored flowers form dense clusters from March to May. Fruits (7–12 cm) [2.75–4.75 in] long) are large and fleshy. Another species of the northern Central Plateau, *Y. torreyi* grows in Tamaulipas, Nuevo Leon, Coahuila, Chihuahua, and Durango.

* * *

Yucca treculeana Carr. is an unbranched columnar form with long leaves and short flowering stalks that elevate the panicle inflorescence only partially above the leaves. Relatively small white flowers (3–5 cm [1.25–2 in] long) are globose or hemispherical in shape and produce small, fleshy, indehiscent fruits. The rigid leaves have a few fine, straight fibers on the margins. Typically flowering in March and April, this is a species of the northern Central Plateau in the states of Durango, Coahuila, Nuevo Leon, and Tamaulipas. It is known by the common name "palma pita" in Coahuila.

BRANCHED YUCCAS

Four of the many species of treelike Yuccas have been selected for discussion: *Y. decipiens*, *Y. filifera*, *Y. jaliscana*, and *Y. periculosa*.

Yucca filifera Chab.
Palma China or Izote

Conspicuous as one of the very large, highly branched species, *Yucca filifera* has pendant, narrowly cylindrical inflorescences with many tightly clustered, creamy white, bell-shaped flowers. Upon pollination, these inflorescences bear fleshy pendant fruits. The long swordlike leaves have brittle, curled filaments along the margins, a

Yucca filifera

TREE to 10 m (33 ft) tall, much branched. LEAVES 50 cm (19.5 in) long, 3.5 cm (1.5 in) wide, filiferous margins. FLOWER 3.5–5.5 cm (1.5–2.25 in) long, creamy-white; inflorescence 1.5 m (5 ft) long, pendant. FRUIT 5–9 cm (2–3.5 in) long, 2.5–3.5 cm (1–1.5 in) in diameter, indehiscent.

condition more readily seen in young leaves. Widely distributed, *Y. filifera* grows most commonly along the east side of the Central Plateau from Coahuila and Nuevo Leon south to Hidalgo and México, with large populations in southern Nuevo Leon and northern San Luis Potosí. It flowers from April to the end of May.

Yucca decipiens Trel. reaches a height of 15 m (50 ft). Leaves are about .5 m (1.5 ft) in length, with many spiral filaments on the margins. Erect inflorescences, well exserted beyond the leaves, form meter-long, conical panicles of numerous flowers. This species, called "palma china," flowers from the end of January to the end of March. It grows in the western portion of the Central Plateau, in the states of San Luis Potosí, Zacatecas, Durango, Jalisco, Guanajuato, and Aguascalientes.

* * *

Yucca jaliscana Trel. may reach 12 m (40 ft) in height, with leaves 1 m (3 ft) long; leaves have fine fibers on their margins. The erect inflorescences barely exceed the leaves. This species flowers during July and August in Jalisco, Colima, and Guanajuato.

* * *

Yucca periculosa Baker, also a much-branched tree form, has short (30 cm [11.75 in] long) leaves with fine filaments along the margins. At the end of each branch a short inflorescence stalk supports the narrowly ovoid panicle of flowers that stands erect above the leaves. A plant of southern Mexico, *Y. periculosa* bears the common name "izote." It can be found in flower in March and April in the states of Puebla, Tlaxcala, Veracruz, and Oaxaca.

Amaryllidaceae—Amaryllis Family

Bomarea hirtella (H.B.K.) Herb.
Granadita or Jícama Montes

Delicate funnel-shaped flowers in summer, followed in the fall by decorative fruits, make *Bomarea* a colorful twining vine. The elliptic-oblanceolate segments of the perianth parts, arising from the top of the sparsely hairy ovary, are pale red, rose-pink, or salmon. They spread about halfway up their funnel-shaped form to reveal three claw-shaped, pale pink or chartreuse inner petals spotted with tiny dark brown dots. The many branched, umbellate inflorescence bears numerous glabrous,

Bomarea hirtella
TWINING VINE. LEAVES to 16 cm (6.25 in) long, 1.5–2.5 cm (.5–1 in) wide, parallel veins. FLOWER 3–3.5 cm (1.25–1.5 in) long; inflorescence umbellate. FRUIT 2–3 cm (.75–1.25 in) broad, turbinate; seeds subglobose.

elongate rays, each with one or more green bracts and, commonly, two or three pendant flowers.

Attractive elongate-lanceolate leaves on short petioles are parallel-veined. Glabrous above, they often are villous or hirtellous (bearing short hairs) below, hence the specific name. The common name "granadita" (little pomegranate) is derived from the fruit, which bursts to reveal a mass of shiny, bright red seeds escaping from a white inner lining. A second common name, "jícama montes" (mountain jícama), refers to the vine's edible, tuberous roots.

Bomarea favors the mountains near Morelia and in Hidalgo, southward to Chiapas and Yucatán. Look for it also around the hills of Mexico City.

Milla biflora Cav.
Estrella

The six-parted perianth on a slender stem arising abruptly from the soil identifies this star-shaped white flower as a member of the Amaryllis Family. The fragrant, waxy corolla, pale greenish on the outside, is marked with prominent darker green median stripes on each segment. The stem bears one or two flowers (hence *biflora*) in a terminal umbel. Six short stamens encircling a longer style stand erect in the center of the wide-open corolla. Several narrow, green, grasslike leaves often wither before the flower appears.

Milla biflora is a common plant from southern Arizona, New Mexico, and Texas in the United States south through the mountains of Mexico into Guatemala. It flowers abundantly on the volcanic slopes near Mexico City from June through September. Antonio Cavanilles, Spanish botanist and director of the botanic gardens of Madrid, honored J. Milla, a court gardener, by naming this plant for him in 1793.

Sprekelia formosissima (L.) Herb.
Flor de Mayo

This large, bright crimson, fleur-de-lis-shaped flower is one of Mexico's most striking native plants. Each flowering stalk, growing

AMARYLLIDACEAE—AMARYLLIS FAMILY

Milla biflora
HERB, perennial; corm 1.5 cm (.75 in) in diameter, brown. LEAVES 1 mm wide, grasslike, basal. FLOWER 5-7 cm (2-2.75 in) in diameter, white. FRUIT capsule with numerous small seeds.

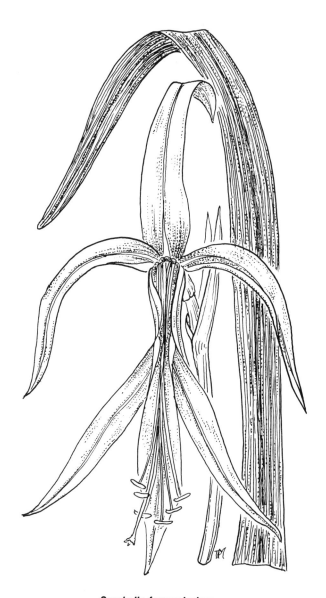

Sprekelia formosissima

BULBOUS HERB 5 cm (2 in) in diameter, tunicate. LEAVES 30–45 cm (12–17 in) long, straplike, linear. FLOWER 8–9 cm (3.25–3.5 in) long, solitary; bract 5 cm (2 in) long, bifid, spathelike; stamens 6; stigma trifid. FRUIT a capsule; seeds flat, black, winged.

individually from a globose bulb, is terminated by a single flower protruding at right angles from a reddish-brown, spathelike, two-parted bract. The upper of the six almost equally sized perianth parts, standing in an erect position, is flanked by two lateral ones, which arch downward. The three lowest roll together for a part of their length to enclose the stamens and style, and then spread outward and downward. Six long stamens of different lengths and a longer style flow downward through the rolled portion of the perianth, gracefully curving over the lowest segment. The capsular fruit contains many flat, black, winged seeds. After flowering, three to six linear, straplike, long leaves develop.

Blooming from April to June, this unusual flower is conspicuous on the Pedregal of Mexico City and grows in the state of Durango south to Guerrero and Oaxaca. It lives up to its specific epithet *formosissima,* which in Latin means "most beautiful."

Anacardiaceae—Cashew Family

Mangifera indica L.
Mango

Producer of a favorite fruit, the mango tree is a beautiful large evergreen with a dense, spreading crown. Young leaves are flaccid, thin, and tan-red, but mature into stiff, dark green, shade-producing masses. The long inflorescence is a terminal panicle with as many as 6,000 reddish, pink, or whitish-green flowers, which are night-blooming and insect-pollinated. Of the numerous flowers in each panicle, only a few bear the female reproductive structures; the others are male. The tree flowers profusely every other year, taking the off-year to restore food reserves depleted by its large crop of fruit. Mango trees are of bearing age between four and six years, increasing in productivity until 20 years old. A large mature tree can produce 2,500 fruits, which within two to five months of pollination are yellow, green, or red, with a smooth, thick skin, and ready to be picked. The flesh is bright yellow-orange and surrounds a single light brown seed in a hard leathery envelope.

ANACARDIACEAE — CASHEW FAMILY

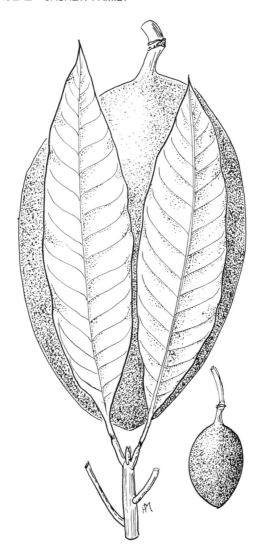

Mangifera indica
TREE 10–40 m (33–130 ft) high; dense spreading crown. LEAVES 10–20 cm (4–8 in) long, oblong lanceolate, alternate. FLOWER 10–15 mm in diameter; inflorescence 10–50 cm (4–20 in) long. FRUIT 5–30 cm (2–12 in) long, ovoid, drupe, edible.

ANACARDIACEAE – CASHEW FAMILY

Mangifera indica, as its specific name implies, is native to India and has been cultivated and naturalized throughout the warmer areas of Mexico. The popular fruit is readily available in local markets.

Pseudosmodingium pernicosum (H.B.K.) Engl.
Cuajiote

Many large panicles, each containing a profusion of small, white, unisexual flowers or masses of lustrous, flattened, small fruits, create an interest in this otherwise undistinguished shrub or tree. Contrasting the white petals of the minute flower is a distinctive dark marking, which outlines the veins of the petal; the network of veins also is prominent on the lower surface of each leaflet. The ovate leaflets of the pinnately compound leaf terminate in rounded apices and contract abruptly into long, extended (acuminate) bases. Each leaflet, light-colored on the upper surface and dark on the lower, is supported by a slender flexible petiolule about half as long as the blade. Numerous fruits—flattened, dark drupes with papery beige wings—are crowded into prominent panicles, which persist for a prolonged period.

Pseudosmodingium grows from Sinaloa southward to Guerrero and Morelos. It belongs to the same plant family as poison ivy, and like that species will produce blistering and eruptions on sensitive skin (*pernicies* in Latin means destructive).

Schinus molle L.
Pirul or Pepper Tree

Schinus is an evergreen tree with a large spreading crown and gracefully drooping branches; the latter bear lacy, pinnately compound leaves with 15 to 17 linear-lanceolate leaflets. It frequently is cultivated as an ornamental shade tree. Small yellowish-white flowers with five oblong petals and ten stamens, borne in terminal panicles, are followed by a profusion of rose-red fruits, which sometimes are used to adulterate pepper, owing to their volatile oil and peppery flavor.

The common name "pirul" probably represents a variation on the name Peru, the native home of *S. molle*. It was introduced into Mexico

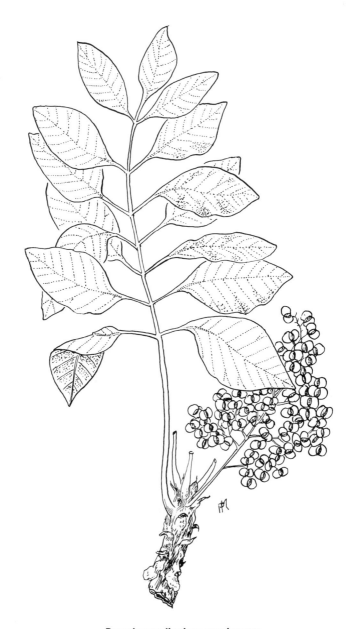

Pseudosmodingium pernicosum

SHRUB or tree 6–20 m (20–65 ft) tall; bark rough, peeling. LEAVES pinnate; 9 to 11 leaflets 3–5.5 cm (1.25–2.25 in) long. FLOWER 1.5 mm long in panicles, white. FRUIT 10 mm wide, 8 mm long, flattened drupe.

ANACARDIACEAE — CASHEW FAMILY

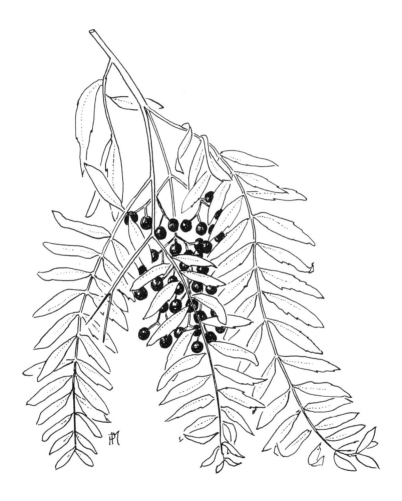

Schinus molle
TREE 10 m (33 ft) tall, evergreen. LEAVES pinnate 15–17 leaflets 4–6 cm (1.5–2.5 in) long. FLOWER small in panicles. FRUIT 5 mm in diameter, rose-red berry.

by Spanish settlers and is now widely distributed throughout the country, by cultivation or through natural establishment. In California it is cultivated under the name "California pepper tree."

Apocynaceae—Dogbane Family

Nerium oleander L.
Laurel Rosa or Oleander

Nerium, a widely recognized shrub or small tree with abundant white, rose-pink, or red-purple flowers, is a native of the Mediterranean region. Terminal clusters of large, funnelform flowers with five united petals and spreading obovate lobes make this an attractive plant; it commonly is used throughout Mexico to line streets and to decorate parks. The spirally overlapping lobes unfurl by twisting open from the bud. Five irregularly cut scales line the throat of the corolla, appearing almost as a second row of petals. Hidden deeper in the flower and tightly surrounding the club-shaped stigma are five stamens. Each flower may produce two slender, follicular fruits, which at maturity open to release numerous seeds, each bearing a tuft of apical hairs. Whorled on short petioles are the leathery, linear-oblong to lanceolate leaves, commonly three or four at a node.

"Oleander" has proved to be exceedingly satisfactory as a cultivated plant because of its ability to remain attractive despite prolonged neglect, drought, and heat. Its principal drawback is the poisonous nature of all its parts. Alkaloids are produced which are sufficiently powerful to have been used in medicine as a cardiac stimulant, although even small amounts may result in death. Allergic reactions can result from contact with the cut stems, and care must be taken that no part of the plant is ingested. Southern Europeans have found this plant to be effective for poisoning both rats and people.

Nerium oleander

SHRUB or tree to 6 m (20 ft) tall. LEAVES 2.5 cm (1 in) long, linear to oblong-lanceolate, whorled. FLOWER 3.5–4.5 cm (1.5–1.75 in) long, funnelform; lobes 2–2.5 cm (.75–1 in) long. FRUIT 10–18 cm (4–7 in) long, slender follicles.

APOCYNACEAE — DOGBANE FAMILY

APOCYNACEAE — DOGBANE FAMILY

Plumeria acutifolia Poir.
Frangipani

In the spring, sweet-scented *Plumeria* blooms appear in crowded, flat, terminal panicles, usually on a leafless tree. The waxy, salverform corolla forms a narrow tube that expands into a yellow throat, then opens into five rounded, white, whorled lobes. From each flower two

Plumeria acutifolia
SHRUB or small tree to 7 m (23 ft) tall, milky sap. LEAVES 5–30 cm (2–12 in) long, in terminal clusters. FLOWER 6–7 cm (2.5–2.75 in) long, white with yellow center. FRUIT follicles diverging at 180 degrees, 10–25 cm (4–10 in) long, 4 cm (1.5 in) wide.

follicles placed end to end develop into pencil-like fruits. Large, dark green leaves appear in clusters near the ends of the stiff, blunt, forking branches at the onset of flowers or thereafter. Numerous prominent parallel veins extending from the midrib to the margin mark the oblong to elliptic leaves, which are pointed at both ends (*acutifolia* = sharp-leaved). Thick, gray-green branches are prominently marked with scars from fallen leaves. The entire plant has abundant, poisonous, milky sap, which is said to produce a good quality of rubber.

Found natively from Baja California, Sonora, and Chihuahua south to Veracruz and Oaxaca, this showy shrub was named for the French botanist Plumier (1646-1706), who was an early traveler in America. *Plumeria* is popularly planted as an ornamental for its fragrant, beautiful flowers. In Aztec times it was highly prized as an addition to the royal gardens.

Stemmadenia palmeri Rose ex Greenman
Berrarco

Stemmadenia has large yellow funnel-shaped flowers, which produce short, broad, paired fruits. Two to five erect flowers are borne together in terminal, reduced racemes. Arising from a conical throat, the corolla consists of five united petals and five equal, spreading lobes, which in the bud are overlapping and twisted (contorted). The peculiar, paired, ovoid follicles are attached and spread at 180 degrees; each is extended into a narrow apex by the persistent style. Together they resemble an old-fashioned waxed moustache. Abundant latex is found in the stems and elliptic leaves of these shrubs or trees.

Stemmadenia palmeri is very similar to *S. tomentosa* and has been considered a variety of the latter. The two differ in the abundance of hairs on the lower surfaces of the leaves: *S. palmeri* has hairs only along and in the axils of the principal veins; *S. tomentosa* has hairs over the entire surface. The two species are reported from Sinaloa, Durango, Jalisco, Nayarit, Morelos, and Veracruz. *S. palmeri* was named to honor one of the great botanical collectors of nineteenth-century western North America, Edward Palmer (1831–1911).

APOCYNACEAE — DOGBANE FAMILY

Stemmadenia palmeri

SHRUB or small tree 2−12 m (6.5−40 ft) tall. LEAVES 6−18 cm (2.5−7 in) long, opposite, elliptic. FLOWER 6−9 cm (2.5−3.5 in) long, yellow. FRUIT paired, each 4.5 cm (1.75 in) long, 3 cm (1.25 in) wide, ovoid.

Thevetia ovata (Cav.) A. DC.
Narciso Amarillo

Large, bright yellow, tubular flowers with spreading obovate lobes, twisted in the bud, attract attention to this shrub or small tree. The few-flowered inflorescences arise as axillary branches. Oblanceolate or oblong leaves have rounded tips and elongated, wedge-shaped (cuneate) bases. The dark green leaves are glabrous above, with a paler, pubescent lower surface on which the midrib and parallel lateral veins are prominent. Peculiar helmet-shaped fruits, broader than long, are fleshy and rose-colored mottled with purple. Enclosed in the hard shell of the fruit are one or two large hard seeds.

Like other species of *Thevetia*, *T. ovata* produces a milky latex and the poisonous glucoside thevetin. The distribution of this plant encompasses the western portion of Mexico from Sinaloa to Chiapas and extends into Guatemala.

Thevetia thevetioides (H.B.K.) K. Schum.
Yoyote

Thevetia thevetioides is a shrub or small tree with alternate, narrow, linear leaves having rolled margins. A poisonous milky latex exudes in small quantities from cut surfaces. Large orange or pinkish-yellow flowers arise singly or in pairs from leaf axils. The corolla of five united petals has a short tube topped by large, showy lobes spirally twisted in bud. Within the open corolla throat are five stamens. The yellow-green fruit develops a fleshy outer layer over a hard inner portion, within which are one or two large, hard, poisonous seeds. These peculiarly shaped fruits commonly are used as rattles in native dances.

Common and frequently cultivated, *T. thevetioides* grows from Michoacán and Tamaulipas south to Veracruz and Oaxaca. It is similar to *T. peruviana,* which differs in having slightly wider leaves, a clearer yellow corolla, and red fruit at maturity. The latter species grows natively in San Luis Potosí, Veracruz, Yucatán, Chiapas, and Guerrero, but cultivation extends its range. All parts of both species contain the glucoside thevetin, which renders them poisonous if eaten.

APOCYNACEAE—DOGBANE FAMILY

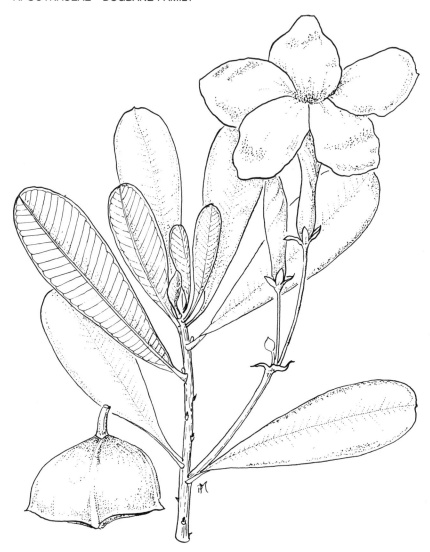

Thevetia ovata

SHRUB or small tree 1–5 m (3–16 ft) tall. LEAVES 6–17 cm (2.5–6.5 in) long, 1–4 cm (.5–1.5 in) wide. FLOWER 8–12 mm broad, yellow; corolla tube 15 mm long; throat 12–15 mm long; lobes 2.5–3.5 cm (1–1.5 in) long. FRUIT 4.5–5 cm (1.75–2 in) long.

APOCYNACEAE—DOGBANE FAMILY

Thevetia thevetioides

SHRUB or small tree 3–9 m (10–30 ft) tall, evergreen. LEAVES 7–14 cm (3–6 in) long, 5–10 mm wide, linear. FLOWER 8–10 cm (3.25–4 in) long, orange or pinkish-yellow. FRUIT 4 cm (1.5 in) wide, 3 cm (1.25 in) long, 3 cm (1.25 in) thick.

Araceae—Arum Family

Monstera deliciosa Liebm.
Ceriman

Monstera deliciosa, the "delicious monster," is aptly named: this epiphytic vine grows rapidly to great size, and its mature fruit is juicy and sweet. The conspicuously large perforated and pinnatifid leaves on long petioles first attract attention to this native of humid tropical forests. In young specimens the immature leaves are cordate, without perforation or dissection. From the entire cordate blade to those having numerous perforations and deep dissections, all gradations can be found. Thick adventitious roots commonly develop from the leaf axils. These tough, ropelike roots are woven into strong baskets.

Small inconspicuous flowers are tightly clustered on a large, fleshy, cylindrical axis (the spadix), which in turn is surrounded by the leathery, pale yellow-green spathe. This inflorescence is supported by a peduncle 1–15 cm (.5–6 in) long. After pollination, the fleshy spadix develops edible, sweet, juicy, whitish or pale yellow segments (berries).

A native of southern Mexico in the states of Veracruz, Morelos, Oaxaca, Guerrero, and Chiapas, *Monstera* extends southward to Panama. It commonly is cultivated as a house or patio plant for its ornamental leaves. In colder climates it is a ubiquitous resident of the tropical greenhouse, although its monstrous growth and tendency to smother other plants has made it the bane of many a greenhouse gardener's existence.

Monstera deliciosa
VINE, large, epiphytic. LEAVES 40–60 cm (16–24 in) long; pinnatifid and perforate; petiole to 1 m (3 ft) long. SPATHE 20–25 cm (8–10 in) long, pale yellow-green. SPADIX 11–20 cm (4.25–8 in) long, pale yellow, FRUITING SPADIX fleshy, juicy, edible.

ARACEAE—ARUM FAMILY

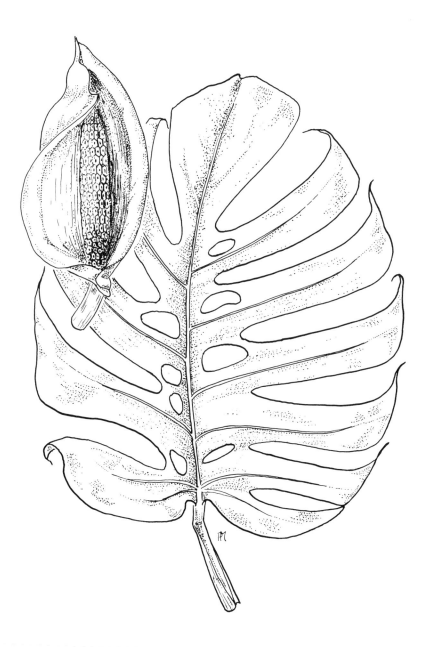

Araliaceae—Ginseng Family

Oreopanax peltatus Linden
Mano de Leon or Tronador

Foliage is the outstanding feature of this highly ornamental tree or shrub, as the flowers and fruits are small and insignificant. Large leaves on long slender petioles are deeply lobed and leathery, yet flexible when fresh. Young foliage usually is stellate-tomentose, becoming glabrous with age. Small flowers are clustered into heads, which form large, many-branched panicles. Fruits are black at maturity. Large old trees may develop a broad, spreading crown.

The leaves (*peltatus* = shield-shaped) are highly prized and suitable for covering baskets or for wrapping food and other articles; in addition to their large size, they are durable, pliable, tough, and do not wilt quickly. Bunches of them often are offered for sale in the markets.

Oreopanax is found in pine and oak forests from Sonora to Chihuahua south to Guatemala. It has been introduced into southern California as a popular ornamental evergreen.

Aristolochiaceae—Pipe Vine Family

Aristolochia grandiflora Swartz
Bonete or Flor de Pato

One of the largest individual flowers in the plant kingdom, "flor de pato" is so named because the huge, unopened buds resemble life-sized ducks suspended from a high-climbing herbaceous vine. Solitary pendant flowers develop from the axils of the leaves, which are ovate to cordate with acute or long-acuminate tips and long petioles. The open flower consists of a long curved tube that flares into a broad oval limb with a slender tail-like appendage. The limb is hairy, yellowish with dark purple spots on the inside, and whitish on the outside. The whole structure resembles an old-fashioned ear trumpet or an oversized smoking pipe.

ARALIACEAE—GINSENG FAMILY

Oreopanax peltatus
SHRUB or tree less than 12 m (40 ft) tall. LEAVES 15–50 cm (6–20 in) long, evergreen, 5–7 lobed. FLOWER small, in heads. FRUIT 6 mm in diameter, black.

ARISTOLOCHIACEAE — PIPE VINE FAMILY

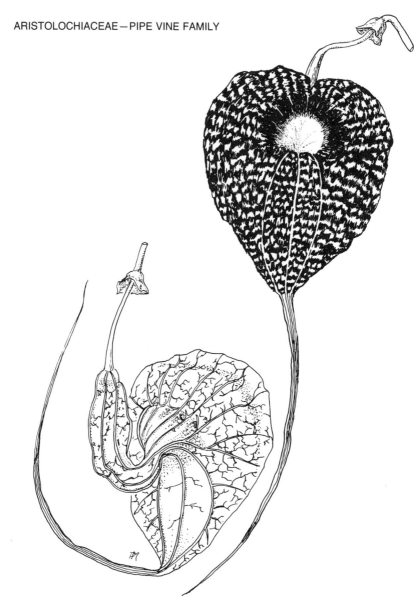

Aristolochia grandiflora
VINE, large, herbaceous. LEAVES 8–25 cm (3–10 in) long, ovate, cordate. FLOWER 12–30 cm (5–12 in) long, limb 15–45 cm (6–17.5 in) broad, tail 1 m (3 ft) long, yellowish-purple. FRUIT 10 cm (4 in) long, 4.5 cm (1.75 in) thick, capsule.

This odd, even grotesque, flower produces a strong repugnant odor, which probably developed to attract specific pollinators. Because of its unusualness, the vine frequently is cultivated in large greenhouses in the United States. "Flor de pato" is found along streams in warm wet areas of Chiapas, Veracruz, and the Yucatán Peninsula.

Asclepiadaceae—Milkweed Family

Asclepias curassavica L.
Hierba de Leche

This common and widely distributed member of the Milkweed Family is showy; bright red flowers with reflexed petals contrast with prominent yellow hoods standing erect in the center of the blossom. Arising from the perennial, underground rootstock are several herbaceous, erect, aerial stems. These bear opposite, or occasionally whorled, lanceolate or linear-lanceolate leaves with short petioles and smooth (entire) margins. From the axils of the upper leaves the few-flowered umbels develop. Narrow spindle-shaped follicles produce many seeds, each bearing a tuft of hairs to make it more easily dispersed by the wind. All parts of the plant exude abundant milky latex when cut or broken.

This milkweed is widely distributed throughout Mexico, the tropics, and the subtropics of the Americas. It grows in moist areas along roadsides, field margins, and stream banks at elevations ranging from sea level to 2000 m (6550 ft). Its place of origin is uncertain, but it was named *curassavica* in 1753 by Linnaeus because the original specimen came from Curacao, Venezuela.

Cryptostegia grandiflora R. Br.
Cuerno

Both the showy, colorful flowers and the large, peculiar fruits make *Cryptostegia grandiflora* a conspicuous woody vine. Clusters of two or three large attractive flowers develop at the ends of each of the many branches. The long funnelform corolla, rose lavender outside and white

ASCLEPIADACEAE—MILKWEED FAMILY

Asclepias curassavica
HERB 60–120 cm (24–48 in) tall; several stems. LEAVES 5–16 cm (2–6.25 in) long, lanceolate. FLOWERS in few-flowered umbel; corolla lobes 5–10 mm long, red; hoods 4–5 mm long, yellow, each with incurved horn. FRUIT 4–10 cm (1.5–4 in) long, follicle narrowly funnelform.

ASCLEPIADACEAE—MILKWEED FAMILY

Cryptostegia grandiflora
WOODY VINE or sprawling shrub; milky latex. LEAVES 4.5–10 cm (2–4 in) long, oval to elliptic-ovate, opposite, petiolate. FLOWER 5–7 cm (2–2.75 in) long, funnelform, pink or purple. FRUIT 12 cm (4.75 in) long, 3.5 cm (1.5 in) thick, paired, angular.

ASCLEPIADACEAE—MILKWEED FAMILY

within, has five widely spreading lobes, which open in a spiral fashion. A calyx of five lanceolate sepals is present at the base of the bloom. This vigorous vine may completely cover a tall tree with its whiplike branches of thick, leathery, glossy, oval leaves. Large three-angled fruits, paired at 180 degrees, are solid and green when young. At maturity they become honey-colored and wrinkled, and split to expose numerous red-brown seeds tipped with a tuft of silky white hairs. The common name "cuerno" (horn) refers to the fruit.

Interest in *C. grandiflora* as a source of rubber developed during World War II. Plantations were established in Haiti, but the difficulty of extracting the milky latex prevented satisfactory yields. The stems also are useful for their fiber content. A native of India, *C. grandiflora* has become naturalized in Sinaloa and other frost-free areas of Mexico where it has escaped from cultivation.

Bignoniaceae—Bignonia Family

Astianthus viminalis (H.B.K.) Baill.
Palo de Agua or Ahuejote

Astianthus is a small graceful tree commonly growing near water, as the Mexican name "palo de agua" suggests. Its preference for sandy or gravelly streambeds, combined with long, narrow, deciduous leaves and linear seed pods, recall the closely related genus *Chilopsis,* the "desert willow" of northern Mexico and southwestern United States (see next entry). In terminal clusters grow numerous large, campanulate, bright yellow flowers with funnelform corollas, the lobes of which are irregular and crinkled. Pendant, linear seed capsules produce numerous small seeds, each with a broad white wing.

Astianthus is common from Colima to Guatemala, and in Puebla and Veracruz.

Astianthus viminalis

TREE 15 m (50 ft) tall. LEAVES 13–18 cm (5–7 in) long, linear, opposite, whorled.▶
FLOWER 5 cm (2 in) long, yellow, funnelform. FRUIT 6–9 cm (2.5–3.5 in) long, 8 mm broad.

BIGNONIACEAE — BIGNONIA FAMILY

BIGNONIACEAE — BIGNONIA FAMILY

Chilopsis linearis (Cav.) Sweet
Mimbre or Desert Willow

The sweet distinctive fragrance of *Chilopsis* flowers often can be detected before the tree can be seen. Attractive white to purplish flowers grow in terminal racemes. A small, two-lipped, inflated calyx subtends the tubular corolla, which consists of five united, irregularly shaped petals, the two upper erect and the three lower forming a platform upon which nectar-seeking insects land. The lobes of the petals are somewhat fringed. Long, pencil-like, light tan pods are filled with numerous flattened seeds bearing silky hairs at each end. This graceful tree has long, narrow, deciduous leaves resembling a true willow of the genus *Salix*, hence the common name "desert willow." Dark brown older bark is irregularly ridged and scaly; new, erect, slender branches are smooth and somewhat striated.

Chilopsis is a feathery tree common in washes and arroyos in the northern states from Baja California to Tamaulipas and south to Zacatecas and Durango. In habitat preference and form, *Chilopsis* is replaced in southern Mexico by the yellow-flowered, closely related *Astianthus viminalis* (see preceding entry).

Crescentia alata H.B.K.
Ayal, Guiro, or Calabash

During the summer months, large exotic flowers develop from nodes along the stem and on the larger branches of *Crescentia*. Later, distinctive large oval or subglobose fruits seem to grow right out of the same trunk. The plant's unique cruciform leaves led the Spanish to believe the tree had religious significance. The entire, evergreen leaf consists of a broadly winged petiole (*alata* = winged) and three linear to obovate leaflets. The unique solitary flowers have five petals united into an oblique, campanulate corolla, greenish purple-brown in color and sometimes streaked with rose-purple. Four prominent stamens and one pistil extend beyond the throat.

BIGNONIACEAE — BIGNONIA FAMILY

Chilopsis linearis
SHRUB or tree to 9 m (30 ft) tall. LEAVES 5–30 cm (2–12 in) long, narrow. FLOWER tubular, 2.5–3.5 cm (1–1.5 in) long, white to purple. FRUIT 10–20 cm (4–8 in) long, pencil-like.

BIGNONIACEAE — BIGNONIA FAMILY

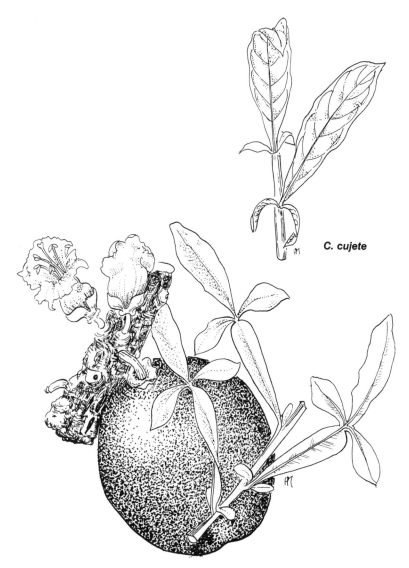

C. cujete

Crescentia alata
TREE to 10 m (33 ft) tall, low spreading. LEAVES shaped like a cross. FLOWER 6–7 cm (2.5–2.75 in) long, greenish purple-brown. FRUIT 10–15 cm (4–6 in) in diameter, baseball size, on trunk and large branches.

Large yellow-green fruits develop a hard outer shell, and, like the flowers, are almost sessile on the trunk and branches. The shells are used for drinking cups, and the sweet pulp is used for various medicinal purposes or in making an alcoholic brew. Tequila often is poured through a hole into the abundant sweet pulp, stirred, and drunk through a straw.

Crescentia is common in places that normally are dry, but tolerates flooded conditions during the rainy season. It is most common along the Pacific slope from Sinaloa to Chiapas. A second species, *C. cujete* L., is found more commonly on the eastern side of Mexico from Tamaulipas to Veracruz and Yucatán, as well as in Guerrero, Oaxaca, and Chiapas. It differs from *C. alata* in having simple leaves, a yellowish-white corolla, and green, somewhat larger fruits.

Distictis buccinatoria (DC.) A. Gentry
Clarín or Trompetilla Grande

Conspicuous because of its brilliant, trumpetlike flowers, *Distictis* is a spectacular large vine that blossoms most of the year. A perfect foil for the red flowers are the glossy evergreen leaves of two oblong or oval leaflets with a third leaflet modified into the branching tendril by which the vine climbs. The inflorescence is a terminal raceme, from which the large, pendant flowers bloom. The color of the corolla grades from red or red-purple lobes, to orange toward the base, to yellow where it meets the short, green, five-toothed calyx. The tube is moderately curved, with spreading, irregular, slightly reflexed lobes. Tan fruit, conspicuously wrinkled on the surface, is narrow-elliptic, tapering obtusely at both ends, with a pattern of five wavy ridges on its surface.

This striking species is found natively from Jalisco and San Luis Potosí south to Puebla. Because *Distictis* makes a brilliant display as it climbs over walls or through trees, it often is cultivated and so is more widely distributed. Several references apply the scientific name *Phaedranthus buccinatorius* to this species.

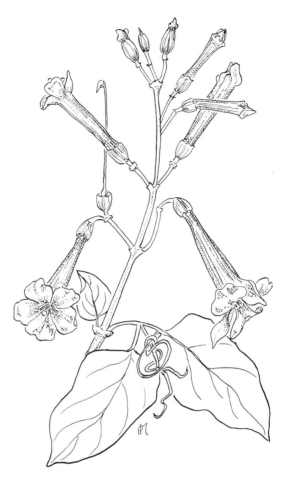

Distictis buccinatoria

VINE woody, climbing by leaf tendrils. LEAVES evergreen; leaflets 5–7.5 cm (2–2.75 in) long. FLOWER 10–15 cm (4–6 in) long, in clusters, red, tubular. FRUIT 14–15 cm (5.5–6 in) long.

Jacaranda mimosifolia D. Don
Jacaranda

Both the filmy leaves and the beautiful pyramidal panicles of lavender-blue blooms make the "jacaranda" tree a favorite in Mexico, although it is native to Argentina. In the spring, before the leaves develop,

BIGNONIACEAE — BIGNONIA FAMILY

this large gray-trunked tree is covered with faintly fragrant flowers, often with more than 80 blossoms in one axillary or terminal cluster. Each flower is made up of a minute cup-shaped calyx and a slender tubular corolla with five united, unequal, delicately fluted petals. The five lobes are rounded, reflexed, and spreading, flaring into a bell shape. Inside the flower are two pairs of fertile stamens, one longer sterile stamen, and a pistil. The flowers quickly wilt and fall in masses after blooming, repeating their color below the spreading branches of the newly leafed tree.

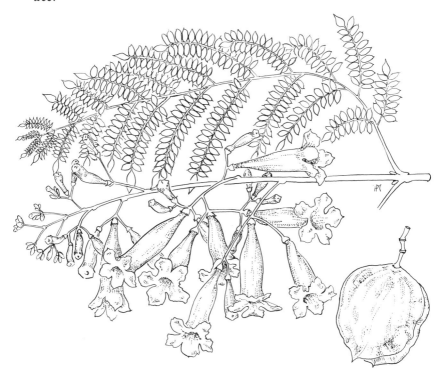

Jacaranda mimosifolia

TREE 8–12 m (26–40 ft) tall, deciduous. LEAVES 20–45 cm (8–18 in) long, 12–20 cm (4.75–8 in) wide, bipinnate; 20–40 paired pinnae; leaflets 9–12 mm long. FLOWER panicle 20–30 cm (8–12 in) long; corolla 3–5 cm (1.25–2 in) long, blue to blue-violet. FRUIT 5–6 cm (2–2.5 in) in diameter.

Graceful fernlike leaves are twice-pinnate, with numerous paired pinnae, each having nine to twenty-two pairs of sessile, oblong to oblong-lanceolate leaflets terminated by a single enlarged one. Each leaflet has an oblique base and a short, pointed tip. The airy, symmetrical foliage makes "jacaranda" an elegant ornamental tree even when not in bloom. Flat, discoid fruits, green when immature, persist on the tree as conspicuous, dark brown, woody capsules with their two sides open like castanet clappers after the thin, dark brown, winged seeds are dispersed. "Jacaranda" is now widely cultivated as an ornamental tree and in some areas has become well established.

Pyrostegia venusta (Ker) Miers
Llamarada or Flame Vine

Masses of flaming orange, tubular, erect flowers covering the ground, a wall, or other plants make *Pyrostegia* one of Mexico's most colorful vines. Each of the individual flowers comprising the terminal or axillary panicles is composed of a slender, tubular corolla of five united, unequal petals. Three of the lobes are strongly recurved, while the remaining two, less deeply divided, stand more or less erect. From the mouth of the corolla protrude two pairs of stamens unequal in length, and a prominent, persistent pistil. The small, green, minutely toothed cup of five united sepals subtending the narrow corolla becomes apparent when the flower tube slips from the calyx to drop or be entrapped as it falls by the enlarging stigma. Bright green, glossy, alternate leaves are pinnately compound. Of the two or three ovate leaflets, one may be modified into a supporting tendril for the climbing vine. The capsular fruit is linear, smooth, and compressed, containing many broad, dark brown, winged seeds.

Although not a native of Mexico, *Pyrostegia* is a popular and handsome (*venusta* = handsome) cultivated plant in warmer areas. Its brilliant crimson-orange masses of flowers and buds, which resemble flames, and its habit of covering whatever it is near gave rise to the scientific name, from the Greek words *pyro* (fire) and *stegia* (covered with or covering).

BIGNONIACEAE — BIGNONIA FAMILY

Pyrostegia venusta

VINE woody, evergreen. LEAVES pinnate; leaflets 4—8 cm (1.5—3 in) long, 2—5 cm (.75—2 in) wide; terminal leaflet a tendril. FLOWER 5.5—7.5 cm (2.25—3 in) long, orange. FRUIT 25 cm (10 in) long, 1—1.5 cm (.5—.75 in) wide; seed 1 cm (.5 in) long, 3.5 cm (1.5 in) wide.

BIGNONIACEAE—BIGNONIA FAMILY

Roseodendron donnell-smithii (Rose) Miranda
Primavera or White Mahogany

Masses of golden trumpet-shaped blossoms transform this nondescript, leafless, light gray tree into a spectacular springtime sight. Showy yellow flowers in large pyramidal panicle clusters appear in May before leaves begin to develop. The tubular calyx is two-lipped, while the tubular corolla has five spreading lobes of unequal form. Fruiting capsules, longitudinally ridged and irregularly tuberculate, are conspicuous by their length and greenish-brown color.

Roseodendron grows from Nayarit and Colima southward to Guatemala. Because of its importance as a lumber tree, in some areas it has been almost eradicated. Many references describe *Roseodendron* as a species of *Tabebuia* or *Cybistax*.

Spathodea campanulata Beauv.
Tulipan de Africa or African Tulip Tree

Fiery red flowers, resembling masses of fringed tulips, crown this handsome tree for a long period in winter and spring. The flowers are startlingly vivid; each bright scarlet lobe is accented and edged with yellow. The terminal clusters are made up of a circular mass of closely packed, horn-shaped flower buds surrounded by an outer circle of opening, bell-shaped blooms. The buds on the outside of the mass open a few at a time and last only two days. Individual flowers resemble a large, irregular, bell-shaped cup with five crispy, frilled, recurving lobes. The scarlet corolla has yellow streaks inside the throat and four projecting yellow stamens with brown anthers.

A peculiar, light brown, horn-shaped calyx splits open on the outer side to release the flower. The name *Spathodea* is derived from this spathelike structure, which is thick-walled, rigid, and velvety. Even when not in bloom this gray-barked upright tree is made handsome by its large, compound, dark green leaves. The leaflets are ovate-lanceolate with abrupt short points, moderately leathery or thickened, the edges slightly rolled under.

This showy African native is cultivated throughout tropical Mexico.

BIGNONIACEAE — BIGNONIA FAMILY

Roseodendron donnell-smithii
TREE to 30 m (100 ft) tall. LEAVES palmate with 5–7 leaflets. FLOWER 5.5–7 cm (2.25–2.75 in) long, tubular, yellow. FRUIT 20–30 cm (8–12 in) long, with ridges; seeds winged.

BIGNONIACEAE—BIGNONIA FAMILY

Spathodea campanulata
TREE to 20 m (65 ft) tall, evergreen. LEAVES opposite pinnate; 9–19 leaflets, each 5–10 cm (2–4 in) long. FLOWER 10–12 cm (4–4.75 in) long, in circular clusters, scarlet. FRUIT boat-shaped 12–25 cm (4.75–10 in) long, dark brown.

BIGNONIACEAE—BIGNONIA FAMILY

Tabebuia chrysantha (Jacq.) Nicholson
Amapa

Flowering during the dry season, especially from January to May, "amapa" is a strikingly beautiful tree covered with large, sweet-scented, bright yellow blooms (*chrysantha* = golden flowered). Condensed in terminal inflorescences, the flowers expand from a tubular base into a broad yellow throat streaked with reddish lines. Three elliptic lobes form the recurving lower lip and two the upper. Four stamens of different lengths and a very short staminodium are attached to the corolla. At the base of the corolla tube are the five united sepals, forming a campanulate calyx clothed in a dense covering of yellowish hairs. An elongated brown capsule, tapering at both ends, is finely striated but may become rough and warty when dry. Many winged seeds are packed inside. Leaves, which appear after flowering, are opposite, palmately compound, with five lanceolate to elliptic leaflets.

The tree is common in both dry and moist tropical forests from southern Sonora into Central America along the Pacific slope, and on the east coast in Veracruz and the Yucatán Peninsula. The dark wood is strong, highly valued, and used for cabinetry or heavy construction.

Tabebuia palmeri Rose
Amapa Prieta

During the winter dry season, *Tabebuia palmeri* becomes a beautiful showy tree, with dense terminal clusters of large pinkish-purple flowers appearing before new leaves are formed. The funnelform corolla expands into spreading lobes, two forming an upper lip and three a lower. A white tomentum covers the exterior corolla lobes, the small campanulate calyx, and the inflorescence branches. The elongated, brown, fruiting capsules are tightly packed with many broadly winged seeds. Shortly after the tree flowers, new leaves develop. They are opposite, palmately compound, with five ovate-oblong or ovate-lanceolate leaflets, somewhat puberulent beneath.

Although first described from a collection made in southern Sonora, this species occurs in the dry woodlands of the Pacific slope as far south as Panama. The dark, reddish-brown wood is very hard, durable, and used for cabinet work, railroad ties, and general construction.

BIGNONIACEAE—BIGNONIA FAMILY

Tabebuia chrysantha

TREE to 30 m (100 ft) tall. LEAVES deciduous, palmate; 5 leaflets 10–18 cm (4–7 in) long. FLOWER 5.5–7.5 cm (2.25–3 in) long, yellow, tubular, fragrant. FRUIT 20–30 cm (8–12 in) long, narrow, rough.

BIGNONIACEAE — BIGNONIA FAMILY

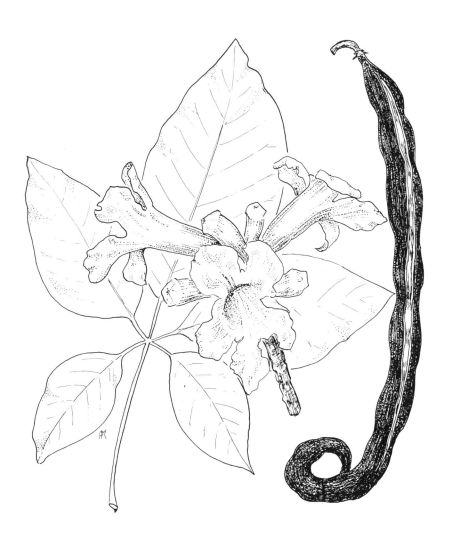

Tabebuia palmeri
TREE 10–15 m (33–50 ft) tall. LEAVES opposite, palmate; 5 leaflets. FLOWER 5–7 cm (2–2.75 in) long, purple, in terminal clusters. FRUIT 35 cm (13.5 in) long, smooth; seeds winged.

BIGNONIACEAE — BIGNONIA FAMILY

Tabebuia rosea (Bertol.) DC.
Palo de Rosa

"Palo de rosa" is one of Mexico's most beautiful trees. Its tall straight trunk, if growing in wet areas, may develop buttresses. Terminating the leafless branches in March or April are open panicles of large, fragile, pink, irregularly shaped flowers. A two-lobed calyx subtends the large, tubular to funnelform corolla, which is white below and expands into five pinkish-lilac to rosy fluted lobes. Three of the lobes form the lower, reflexed lip; the other two become the upper lip. Four functional stamens of two distinct lengths, plus one very short staminodium, are enclosed within the corolla. Dense glandular scales cover the elongated capsular fruit, which contains numerous large winged seeds.

Leaves reappear at the end of the dry season, which terminates flowering. Opposite, palmately compound leaves on long petioles bear five obovate to elliptic leaflets of irregular size. The two lower leaflets, radiating from a common center, are small on short stalks (petiolules); the other three, considerably larger, have supporting petiolules 6 cm (2.5 in) long.

Tabebuia rosea is widely distributed throughout the tropical deciduous forests from southern Sinaloa and Tamaulipas along both coasts to southern Mexico and Central America. It often is cultivated as an ornamental. It also is an important source of lumber, prized for heavy construction. Because of its color and marking, the hard, durable wood may be used for cabinet veneer and interior finishing, for it takes a fine polish.

Tecoma stans (L.) Juss.
Retama or San Pedro

One of the delicate, graceful plants of the hillsides and roadsides of Mexico, *Tecoma stans* displays clusters of bright yellow, fragile flowers. The curved, tubular bud opens to reveal a five-lobed, bilabiate, campanulate blossom with softly recurving lobes, as is typical of the Bignoniaceae. Four stamens are barely as long as the tube. In graceful, terminal racemes, these clusters of little trumpets add color to the land-

BIGNONIACEAE—BIGNONIA FAMILY

Tabebuia rosea
TREE to 25 m (80 ft) tall. LEAVES opposite, 10–35 cm (4–13.5 in) long including petiole, palmate; 5 leaflets, 2 lower 3–8 cm (1.25–3 in) long; others 7.5–16 cm (3–6 in) long. FLOWER 7–10 cm (2.75–4 in) long, pink, tubular. FRUIT 3.5 cm (1.5 in) long; seeds 2–3 cm (.75–1.25 in) long, winged.

BIGNONIACEAE — BIGNONIA FAMILY

Tecoma stans

SHRUB or small tree to 7 m (23 ft) tall. LEAVES opposite, pinnate; leaflets 5–13. FLOWER 3.5–5 cm (1.5–2 in) long, tubular, bright yellow. FRUIT 10–20 cm (4–8 in) long, elongated pod; winged seeds.

scape from autumn to spring. Flattened, linear seed capsules contain many winged seeds.

The foliage of *Tecoma* is as graceful as its flowers. Leaflets of the opposite, pinnate leaves are lanceolate to ovate-lanceolate, with delicate serrated margins. Usually glabrous and yellow-green, thin, and flexible, they add to the beauty of this plant, which is widely distributed and cultivated throughout Mexico. Because it attracts attention both as a wild and a cultivated plant, 34 common names are recorded for this handsome shrub.

Bixaceae—Arnotto Family

Bixa orellana L.
Arnotto or Achiote

Bixa is attractive in the late fall when conspicuous dainty pink flowers form in terminal panicle clusters or when the unusual fruits are present. Each large fragrant flower, with its five showy, separate petals, has abundant, prominent, lavender stamens. In bud the corolla is covered by five brownish-green sepals with reddish-brown scales; this calyx falls off soon after the flower opens. The large, evergreen, alternate leaves on long slender petioles make an attractive background for the flowers and fruits. The leaf is ovate, with an extended narrow tip and a cordate base. Ornamental, reddish-brown fruiting capsules, covered with long soft bristles, contain numerous seeds surrounded by an orange-red pulp. An orange sap is found in the inner bark.

"Arnotto," an orange-red dye, is obtained by heating the seeds in cooking oil or boiling water. It is used to color margarine, butter, cheese, and other foods, and as a dye for fabrics and varnishes. The Indians have used it to paint their bodies for ornament and as protection against mosquitoes. The specific epithet *orellana* commemorates Don Francisco Orellana, famous comrade of Pizarro and early explorer of the Amazon from its source to its mouth. *Bixa* is found in tropical areas from Sinaloa to Chiapas and in Veracruz and Yucatán.

Bixa orellana

SHRUB or small evergreen tree to 5 m (16 ft) tall. LEAVES 9–25 cm (3.5–10 in) long; petioles 2.5–7.5 cm (1–3 in) long. FLOWER 5 cm (2 in) in diameter, pink, in terminal clusters. FRUIT 2.5–4 cm (1–1.5 in) long, reddish-brown with soft prickles.

Bombacaceae—Bombax Family

Bombax ellipticum H.B.K.
Amapola or Clavellina

Large, showy, feathery flowers appear singly or in pairs in late winter or early spring on this leafless tree. Growing upright from the bare branch, the tubular, brownish-purple calyx forms a persistent cup 1.5 to 2 cm deep, from which the purplish bud elongates. Upon maturity the corolla splits into five linear, recurving, narrow petals, which are minutely hairy, purplish-brown on the outside, and pinkish-white within. Conspicuous parts of this flower are the numerous, long, pink stamens fused into a short tube at the base, which give the flower the appearance of a bright pink or white brush, or long-tasseled pompon. A large, dark-brown capsule contains numerous seeds embedded in a mass of silky white hairs, which has the quality of kapok when used to stuff cushions and mattresses.

Bombax ellipticum is a thick-trunked, spineless tree with smooth (or in older specimens, deeply furrowed) green or gray bark. The foliage consists of large, glabrous, palmately compound leaves with five broadly obovate to elliptic leaflets. Young leaves are conspicuously reddish, turning green with age. This species occurs along both the Atlantic and Pacific slopes from Tamaulipas and San Luis Potosí to the Yucatán Peninsula on the east, and from Baja California and Sinaloa to Chiapas on the west.

Bombax palmeri S. Wats.
Clavellina

"Clavellina" is similar to *Bombax ellipticum*, but differs in the palmately compound leaves composed of tomentose leaflets with pointed apices. It blooms in the spring while the tree is leafless. Spectacular flowers, with five whitish, linear petals, have numerous, showy, long, pink to purple stamens fused at the base into a short tube. The petals are longer in relation to the cluster of stamens than in *B. ellipticum*. As the corolla splits open, the petals curl back, allowing the conspicuous stamens to burst into a large, open, colorful tassel. The oblong or

BOMBACACEAE—BOMBAX FAMILY

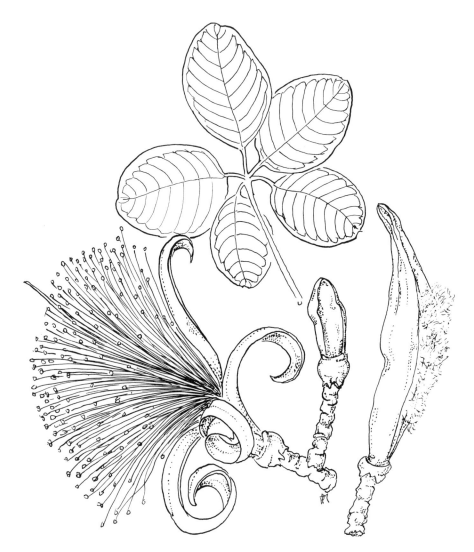

Bombax ellipticum

TREE to 30 m (100 ft) tall, unarmed. LEAVES palmate; 5 leaflets 10–25 cm (4–10 in) long, oval to ovate-elliptic. FLOWER 7–13 cm (2.75–5 in) long; stamens 7–13 cm (2.75–5 in) long, showy, numerous. FRUIT 15–25 cm (5.75–10 in) long, seeds in "dirty" white cotton.

Bombax palmeri

TREE 8–10 m (26–33 ft) tall, unarmed. LEAVES palmate; 5 leaflets 10–25 cm (4–10 in) long, obovate to orbicular, acute at apex, tomentose. FLOWER 10–17 cm (4–6.5 in) long, white; stamens 10–12 cm (4–4.75 in) long. FRUIT 10–12 cm (4–4.75 in) long; seeds numerous, embedded in brownish cotton.

ellipsoid fruit has numerous seeds embedded in brownish cotton. *B. palmeri* is distinguishable by the more abundant hairs on its leaves, even in age, by its pointed leaflet tips, by its darker stamens, and by the darker brownish hairs around its seeds. All are comparative characters and are difficult to recognize when only one specimen is available. The common name "clavellina" refers to the color of the stamens, *clavel* meaning pink in Spanish.

This species is found along the Pacific slope from Sonora to Jalisco.

Ceiba aesculifolia (H.B.K.) Britten & Baker
Pochote

This deciduous ornamental shade tree is very similar to *Ceiba pentandra* (see next entry), but is smaller and has broader, elliptic leaflets, larger flowers, and silky brown hair in the capsule. Stout, conical spines arm the thick trunk and young growth. The alternate leaves on long, slender, finely hairy petioles have five to seven glabrous leaflets, which are whitish-green beneath. Blooming in late winter, the large white flower on a short stout peduncle has a bell-shaped calyx beneath the yellowish hairy corolla, which opens into five spreading and recurving, oblong petals; these turn brown with age. Exceeding the length of the petals are numerous conspicuous, purple-red or white stamens fused into five clusters.

The seed capsule splits along five fine lines to reveal many small seeds surrounded by a mass of brownish, cottony hairs. The silky fibers are used like kapok, and in some cases are superior to it, especially for insulating. Raising *C. aesculifolia* for the fibers has become an important industry in Guatemala. The distribution of this species in Mexico is limited to the west coast from southern Sonora to Chiapas.

Ceiba pentandra (L.) Gaertn.
Pochote or Kapok

This enormous tree of the Bombax Family, even when not in bloom, will draw attention by its big balls of whitish tufts hanging from persistent seed pods. The fruit is a woody capsule, which splits along five lines

BOMBACACEAE—BOMBAX FAMILY

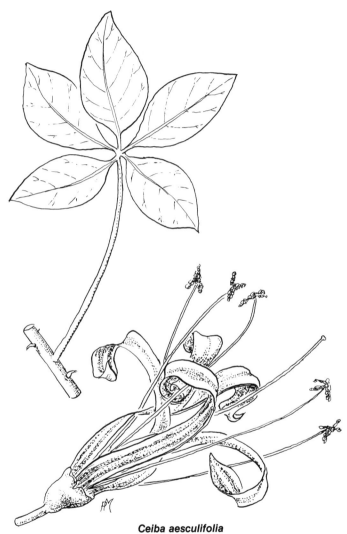

Ceiba aesculifolia
TREE to 25 m (80 ft) tall; conical spines on trunk and young branches. LEAVES 5–15 cm (2–6 in) long, palmate; leaflets 5–7 cm (2.–2.75 in) long. FLOWER 10–16 cm (4–6.25 in) long, white; stamens fused into 5 clusters. FRUIT 12–18 cm (4.75–7 in) long; seeds in brownish cotton.

BOMBACACEAE—BOMBAX FAMILY

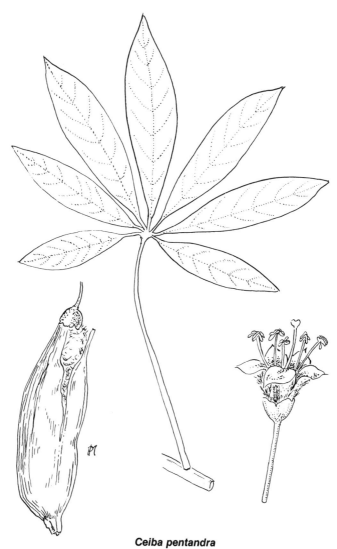

Ceiba pentandra

TREE 40–50 m (130–160 ft) tall, often with buttresses; conical spines on young parts. LEAVES palmate; 5–7 leaflets 8–20 cm (3.25–8 in) long. FLOWER 3–3.5 cm (1.25–1.5 in) long; stamens fused into 5 clusters. FRUIT 10–12 cm (4–4.75 in) long; seeds in grayish cotton.

to expose a dense mass of useful, woolly hairs (kapok), in which black seeds are embedded. While the tree is leafless, relatively small fascicles of fragrant white to pink flowers with dense silky brown hairs on the outer surface bloom. Numerous stamens are fused into five bundles resembling five simple stamens with large, lobed anthers (*pentandra* = five stamens), which in turn are fused into a short tube at the base.

Large, narrow buttresses help support this huge tree. Mature trunks have a smooth gray to gray-green bark, but young stems may be armed with stout, conical spines. The leaves are palmately compound, with five to seven oblanceolate leaflets.

The silky cotton of the fruit is used as a stuffing for mattresses and life preservers, and as insulating material. Seeds produce an oil used for illumination and in producing soap. In Mayan countries the tree had a religious significance and rarely was cut.

Ceiba is found in tropical regions along both the Atlantic and Pacific slopes, from Tamaulipas to the Yucatán Peninsula on the east and from southern Sonora and Baja California to Chiapas on the west, continuing to South America. It often is planted around homes and has been widely used horticulturally in Asia and Africa.

Boraginaceae—Forget-me-not Family

Cordia boissieri DC.
Apacahuite or Trompillo

Delicate white tubular flowers appear scattered over this shrub or small tree as they grow in branched, curled terminal inflorescences. Of the many buds in a cluster, usually only one to four open at the same time, thus extending the flowering period. The wavy white corolla, which turns brownish with age, has five spreading lobes and a yellow throat. A yellow-greenish cylindrical calyx with five short teeth persists, partially covering the maturing fruit. An attractive, ovoid fruit is light yellow, but the shiny surface becomes reddish-brown as it matures,

Cordia boissieri

SHRUB or small tree to 8 m (26 ft) tall. LEAVES to 20 cm (8 in) long, 15 cm (6 in) broad, scabrous. FLOWER 3–4.5 cm (1.25–2 in) long, white, showy. FRUIT 2.5–3 cm (1–1.25 in) long, ovoid.

with a sweet flesh covering the single large seed. It is edible, but there are reports that eating too many will result in dizziness. Leaves of *Cordia* are ovate to oblong, variable in size, dark gray-green, scabrous (rough) on the upper surface, and velvety light green beneath. The bark is thick, gray, and ridged.

This attractive plant is found natively in Coahuila, Nuevo Leon, Tamaulipas, and San Luis Potosí. It frequently is cultivated in other areas. Throughout Mexico there are 30 native species of *Cordia*, ranging in size from small shrubs to tall trees reaching 20 m (65 ft) in height. They all have white or yellow flowers and fleshy, drupaceous fruit.

Burseraceae—Bursera Family

Bursera fagaroides (H.B.K.) Engler
Cuajiote Amarillo or Bursera

The blue-green or yellow-green bark of *Bursera fagaroides* eventually peels from the short trunk and larger branches in straw-colored, parchmentlike sheets. Size, shape, and number of leaflets vary greatly in the alternate, pinnately compound leaves; based on this variation, botanists have recognized three varieties in this species. Leaflets vary in number from three to eleven and in shape from elliptic to narrowly lanceolate. Although cut branchlets are sparingly resinous, they do not produce the strong fragrance typical of many other species of *Bursera*. Flowers, which form the short sessile inflorescences that appear at nodes, are few, small, and inconspicuous; from these develop one to three drupelike, three-angled fruits, each with one seed.

This species is widely distributed from Sonora, Chihuahua, and Coahuila south to Morelos, Guerrero, and Veracruz. Some *Bursera* produce a gum known as "copal," which is burned as incense; some are used as living fence posts from rooted branches. All have small, inconspicuous flowers and two- or three-angled drupelike fruits.

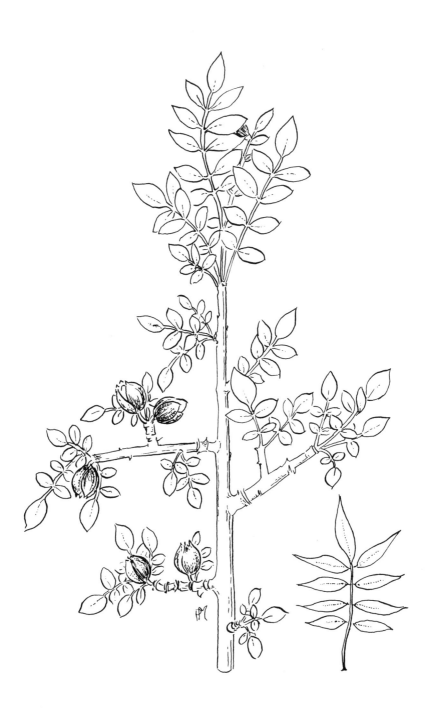

Cactaceae—Cactus Family

This diverse group has successfully adapted itself for survival in arid regions by developing water-storage cells and a thick cuticle, by doing without leaves in many cases, and by protecting itself with spines or poisonous alkaloids. These plants are a botanical family by virtue of their flower structure, not because of their external form.

There are a number of misconceptions about cacti: (1) "All spiny plants are cacti." A number of plants, such as ocotillo and agave, have developed spines, but they belong to families other than the Cactus Family. (2) "All fleshy, succulent plants are cacti." The fleshy structure has developed many times in the plant kingdom in response to aridity; succulent representatives are found among the yuccas, agaves, euphorbias, milkweeds, and many others. (3) "All cacti live in desert areas." Several species of cacti are adapted to life in humid tropical forests, where epiphytes and terrestrial forms use neighboring tree trunks and branches for support.

Members of the Cactus Family have developed in a wide range of sizes and shapes, from small "buttons" only a few centimeters above the ground to massive, multiple-branched candelabra trees several meters tall and weighing thousands of kilograms. While most do not have leaves, or have merely ephemeral ones, some tree or shrub members of this family produce broad deciduous leaves.

Along the stems of all cacti are areoles or nodes from which leaves, flowers, branches, adventitious roots, or spines develop. Spines have at various times been considered as branches, as structures without homology, or as modified leaves. They may be few or many, large, small, or absent. In addition to or in place of large spines, some cacti may develop tufts of hairs; in the chollas, prickly pears, and their close relatives, numerous small, barbed bristles called glochids fill the areoles along with the spines.

Despite all the vegetative variation, the basic flower structure re-

Bursera fagaroides

◀ SHRUB or small tree to 10 m (33 ft) tall; bark smooth, blue-green or yellow-green. LEAVES 3–6 cm (1.25–2.5 in) long, pinnate; 3–11 leaflets. FLOWER inconspicuous. FRUIT 5–6 mm long, drupe, three-angled.

mains the same. The inferior ovary consists of one cell with numerous ovules from which seeds form. The fruit at maturity may be fleshy and edible or hard and dry; often it is spiny.

Sepals and petals arise from the top of a long or short floral tube. A gradual transition can be seen from the green outer sepals to brightly colored inner petals. Numerous stamens line the inside of the floral tube; in the very center of the flower is the style, with three to many stigmas.

The Cactus Family, which with one exception grows natively only in the Americas, is one of the largest and most diverse groups of plants in Mexico. More than a thousand species are recognized by botanists; readers with a particular interest in cacti are referred to *Las Cactaceas de Mexico* by Helia Bravo Hollis (in Spanish). Only a few of the larger, more common, or widely distributed species are described in this handbook.

Carnegiea gigantea (Engelm.) Britt. and Rose
Saguaro

"Saguaro" is a treelike cactus encountered by the traveler in Arizona before entering Mexico and is found throughout the desert areas of Sonora. These massive plants remain unbranched while young, but may develop up to 20 arms as they mature. The 12 to 30 prominent ribs on the main stem and each of the branches have regularly spaced areoles containing 15 to 30 spines. During April and May the stems and branches are crowned with numerous, white, waxy flowers. In June and July the scaly green mature fruits split along three lines, recurving to expose the red inner lining and a multitude of black seeds. To the uninitiated, the "saguaro" appears to be blooming a second time, with red rather than white flowers.

"Saguaro," the state flower of Arizona, played an important role in the life of the early inhabitants of Sonora and Arizona: fruits were

Carnegiea gigantea
COLUMNAR 15 m (50 ft) tall, branched or unbranched. FLOWER 5 cm (2 in) ▶ diameter, 10–12 cm (4–4.75 in) long, white, at apex. FRUIT 5–7 cm (2–2.75 in) long, red inside; seeds black.

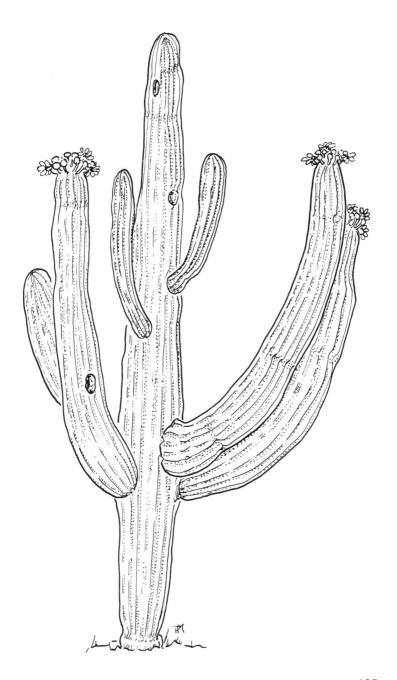

eaten or were fermented for beverage; seeds, with their rich supply of fats, were gathered to enrich the desert diet; woody ribs from dead "saguaros" were used for shelters, roof supports, and corrals. Animals and birds of the desert also depend upon this giant cactus for food and shelter. A procession of birds, including the elf owl, make use of a hole first burrowed in the fleshy stem by the Gila woodpecker.

Echinocactus grandis Rose
Bisnaga or Barrel Cactus

The common name "bisnaga" is applied to many barrel-type cacti of Mexico, but *Echinocactus grandis* is one of the largest members of the genus. Its many narrow ribs bear areoles that are separate at first but continue to grow, often joining as the plant ages. Several very short spines grow from the areoles; the spines at first are yellow, but soon turn reddish-brown. Five or six are radial spines surrounding one erect, straight, central spine. Yellow flowers, embedded in woolly hairs derived from felty young areoles and the axillary wool of the scales of the ovary, crown this giant barrel. The upper scales of the ovary are rigid and spine-tipped; these grade into the broader ovate perianth segments. Mature fruit remains well embedded in the same axillary wool that surrounded the flowers.

Although its distribution is restricted to the limestone hills of Puebla, where it is abundant, this species is a representative of several barrel-type cacti found in desert areas throughout the country.

Epiphyllum oxypetalum (DC.) Haworth
Dama de Noche

This widely cultivated, terrestrial or epiphytic plant of the moist tropical forests is very uncactus-like. The broad basal portion of the stem is round, elongated, and branches into a flattened, almost leaflike structure with crenate margins. Within the notches of the lobes are buds that may further develop, either into more branches or into flowers. If the latter occurs, the flower bud elongates, producing a long narrow

CACTACEAE—CACTUS FAMILY

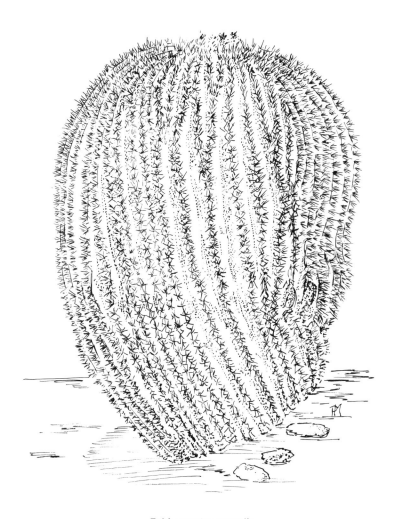

Echinocactus grandis

BARREL unbranched 1−2 m (3.25−6.5 ft) tall, 60−100 cm (23.5−40) in) in diameter. SPINES stout; 5−6 radial 3−4 cm (1.25−1.5 in) long; 1 central. 4−5 cm (1.5−2 in) long. FLOWER 4−5 cm (1.5−2 in) long in woolly hairs, yellow. FRUIT 4−5 cm (1.5−2 in) long.

CACTACEAE—CACTUS FAMILY

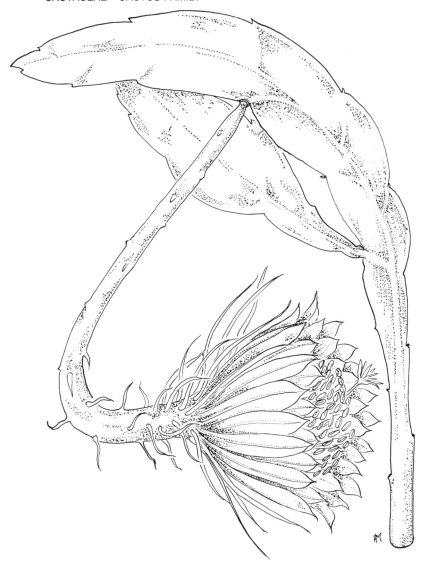

Epiphyllum oxypetalum
EPIPHYTE, stems 3+ m (10+ ft) long. BRANCHES flat 10–12 cm (4–4.75 in) wide. FLOWER 25–30 cm (10–11.75 in) long.

CACTACEAE — CACTUS FAMILY

tube, reddish in color, about 1 cm in diameter, and bearing widely spaced scales. At the apex of the tube are numerous perianth segments, reddish or amber on the outside, grading into white among the inner ones. Opening at night, the flower exposes many graceful white stamens and a thick style. The blossom is so long it tends to hang and bend into a J-shaped structure, opening the throat to night-flying, long-tongued, nectar-seeking pollinators. Oblong red or purple fruits split at maturity to reveal many black seeds surrounded by a white pulp.

The name *Epiphyllum* is derived from the Greek (*epi* = upon, *phyllum* = leaf) and alludes to the flowers being borne on a leaf. This is a misnomer; the flowers, like those of all cacti, are borne on a stem, which in this species is flattened and leaflike. *Epiphyllum* has no leaves, nor does it produce spines. *Oxypetalum* translates as "sharp petals."

Veracruz, Chiapas, and Oaxaca form the northern boundary of this species, which grows throughout Central America and south to the tropical forest of Brazil.

Myrtillocactus geometrizans (Mart.) Console
Garumbullo

The short but well developed trunk of this arborescent form supports many slightly curving branches, which imparts the overall appearance of a large cup. Woolly areoles at intervals of 1.5 to 3 cm on the crests of the ribs produce the very short radial spines that surround the large, daggerlike central spine. Diminutive diurnal flowers have but a few perianth parts. Numerous stamens, well exserted when the flowers are fully open, surround a style tipped with three to five stigmatic lobes. Small tuberculate, globose fruits, brownish-purple upon maturity, are eagerly sought for their spinelessness and flavor. They are eaten fresh, made into refreshing drinks and marmalades, or may be dried like raisins. In the markets they are sold under the name "garumbullo."

Found from Tamaulipas to Guerrero and Oaxaca, this species is abundant in the states of Queretaro, Hidalgo, Guanajuato, and San Luis Potosí. In the west, it is found in Durango, Zacatecas, Jalisco, and Michoacán.

CACTACEAE—CACTUS FAMILY

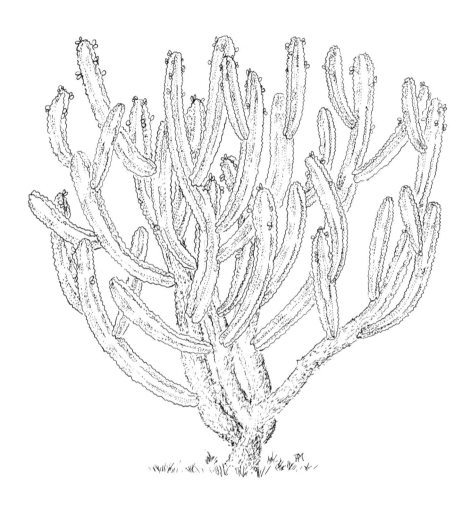

Myrtillocactus geometrizans
COLUMNAR short-trunked, 4 m (13 ft) tall; branches many, 5-ribbed. SPINES radial 2–10 mm long; central 7 cm (2.75 in) long. FLOWER 2.5–3.5 cm (1–1.5 in) long, day-blooming. FRUIT 1–2 cm (.5–.75 in) in diameter, tuberculate.

Neobuxbaumia mezcalaensis (Bravo) Backeberg

Despite its wide distribution, this species of columnar cactus only recently has been recognized as a distinct species. The tall, unbranched stem has 15 or more prominent ribs, which produce separate, closely spaced, yellow, felty areoles. Each areole has five to nine moderately short radial spines and usually one (but occasionally four) central spine. Funnel-shaped, greenish-white to reddish-purple, night-blooming flowers develop in the areoles along the stems. The tuberculate, globose, or pear-shaped fruit opens into an irregular star-shaped form to expose numerous black seeds embedded in white pulp.

Populations of this columnar cactus can be found from Jalisco and Colima south to Oaxaca, as well as in Morelos and Puebla.

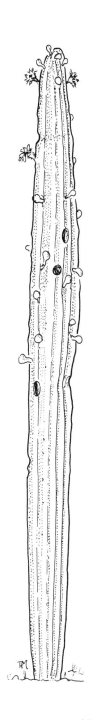

Neobuxbaumia mezcalaensis
COLUMNAR 5–10 m (16–32 ft) tall, unbranched. SPINES 5–9 radial 8–20 mm long, 1 central; areoles separated. FLOWER 5 cm (2 in) long, develop along stem, night-blooming. FRUIT 5 cm (2 in) in diameter, globose.

CACTACEAE—CACTUS FAMILY

OPUNTIA SPECIES

Opuntia, a large and ubiquitous genus, can be divided into two groups: the prickly pears, with flat-jointed stems, and the chollas, with round-jointed stems. Although long, conspicuous spines may or may not be present, glochids—those tiny barbed spines so difficult to see—are present in all species. The flat-stemmed forms range from large, branched, treelike plants reaching 5+ m (16+ ft) in height to small prostrate plants only a few centimeters tall. Flower color is variable, ranging from yellow through orange to cerise.

In some species the ripened ovaries are dry and inedible, but other species are cultivated for their juicy, flavorful fruit, which is eaten raw or made into juice or jams. Care must be exercised when handling the fruits of uncultivated specimens to avoid painful contact with the numerous glochids. The flat pads also are eaten, raw or cooked; both pads and fruits are found for sale in the local markets throughout Mexico. It is with *Opuntia* that Luther Burbank worked to select spineless forms in an attempt to develop food for humans and cattle. Although specifically American in origin, several species of the flat-stemmed opuntias now are worldwide in distribution; *O. ficus-indica* commonly is cultivated in Mediterranean countries for its succulent, edible products.

The cholla type of *Opuntia* ranges from forms with a definite trunk that may reach 2 m (6.5 ft) in height to small bushy or prostrate forms only a few centimeters tall. Flower color, like that of the prickly pears, is variable, even to displaying a wide range of color variation among individuals of the same species. Hues of red, orange, yellow—even purple—with varying amounts of green have been noted. Unlike the prickly pears, the fruits are not considered edible, as they usually are dry or at least lack sufficient juice to make them palatable.

Within this group are certain species to which the name "jumping cholla" has been applied. Cacti do not "jump," but the joints may be detached so readily that a person can become impaled on the large barbed spines without realizing contact has been made. This easy disjunction is an adaptation for dispersal; detached segments easily root to start a new plant in a new location.

Although humans do not, as a rule, eat cholla, cattle raised in the desert turn to the cactus when food is scarce and frequently can be seen with joints of cholla clinging to their faces. A satisfactory way to remove joints is to slip a comb under the spines and lift, thus preventing the joint from rolling to impale one again.

Opuntia has featured importantly in the ancient and modern history of Mexico. Migrating Aztecs (1325 A.D.) were instructed by an oracle to establish their city where they found, perched on a stem of prickly pear, an eagle with a serpent in its talons. The location of Tenochtitlan, now Mexico City, was thus determined. This legend has been perpetuated by the symbolic design in the center of the Mexican flag and on the official coat-of-arms. Of the more than 104 recognized species of *Opuntia* in Mexico, two representative forms are detailed here.

Opuntia ficus-indica (L.) Mill.
Nopal de Castilla or Prickly Pear

This species of *Opuntia* is probably the most widely distributed of any of the cacti; its worldwide acceptance and cultivation resulted from its edible fruits and pads, which are free of large spines. Growing as a tree-form with large oblong or obovate segments, it consists of pads with widely spaced areoles, from which protrude soft yellow glochids, which soon drop off. Abundant showy yellow flowers originate from the areoles on the margins of the flattened pad. A cup-shaped flower consists of many green sepals grading to numerous yellow petals, which enclose a multitude of stamens; this entire structure arises from the top of a tuberculate ovary. At maturity the ovary enlarges, becomes smooth, and changes from green to a red or purple, fleshy, juicy, edible fruit ("tuna").

Because this species of *Opuntia* was cultivated before the Spanish conquest its native home is not known, although it is thought to be in Mexico. It is commonly cultivated on the high plateau of Central Mexico, as well as in the states of México, Puebla, and Oaxaca.

Opuntia imbricata (Haworth) DC.
Abrojo or Tree Cholla

This widely distributed tree-type cholla represents the cylindrical species of *Opuntia*. It develops a short but definite trunk and freely

CACTACEAE — CACTUS FAMILY

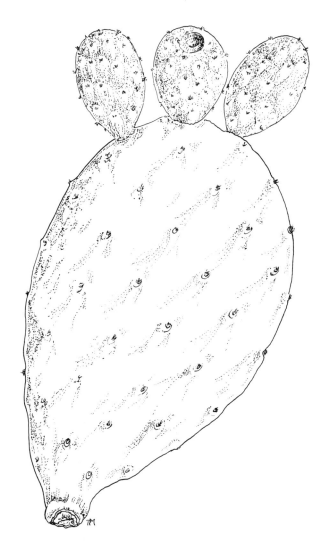

Opuntia ficus-indica
TREE 3–5 m (10–16 ft) tall; trunk definite; joints flat, oblong, 30–60 cm (12–24 in) long. AREOLES distant; spines few or absent; glochids present. FLOWER 6–8 cm (2.5–3 in) long, 8–10 cm (3–4 in) broad. FRUIT 5–10 cm (2–4 in) long, 4–8 cm (1.5–3 in) in diameter.

CACTACEAE—CACTUS FAMILY

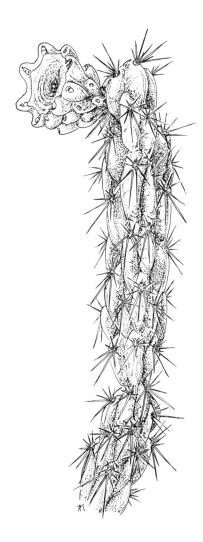

Opuntia imbricata
TREE 1–2 m (3–6.5 ft) tall; joints round, 12.5–38 cm (5–14.75 in) long, tuberculate. SPINES 1.5–3 cm (.75–1.25 in) long; 10–30 per areole. FLOWER 5–7 cm (2–2.75 in) in diameter, 2.5–3.5 cm (1–1.5 in) long.

branches into a dense crown. The cylindrical joints have numerous prominent tubercles, each single yet long enough to overlap those to the right and left, hence the name *imbricata*. At the apical end of each tubercle is an areole with short but numerous spines and abundant glochids. Ephemeral, narrowly conical leaves about 15 mm long develop on new growth. At the ends of the terminal joints, new growth may appear from among clusters of purple flowers. Numerous perianth parts, grading from green sepals to purple petals, surround the many purplish-green stamens. A green, tuberculate ovary develops into a rough yellow fruit, which lacks spines but bears glochids. Although the fruit is slightly fleshy, it is neither edible nor palatable.

Abundant on the high Central Plateau, *Opuntia imbricata* grows in Distrito Federal and the state of Hidalgo northward to the border states of Tamaulipas, Coahuila, and Chihuahua.

* * *

Pachycereus pectin-aboriginum (Engelm.) Britt. & Rose
Cardón

"Cardón," a name applied throughout Mexico to several of the larger species of cactus, is here used to designate a tree with many substantial branches arising above a definite trunk. The 10 to 12 ribs have closely crowded areoles bearing 8 to 12 relatively short spines. Beginning in January and continuing through spring, numerous white flowers are produced on the top meter (3 ft) of the stems. Very spiny yellow fruits develop during early summer. Under all the spines of this formidable burr is an edible, fleshy pulp, which can be made into a preserve, and numerous black seeds, which yield an oily paste or mash.

This species is common on the Pacific coastal slope from Sonora and Baja California south to Oaxaca.

Pachycereus pectin-aboriginum
TREE 5–10 m (16–32 ft) tall, massive; many branches; definite trunk ¼ to ½ ▶ height of tree. FLOWER 5–7.5 cm (2–2.75 in) long, along stem. FRUIT 6–7.5 cm (2.5–3 in) long, yellow, spiny burr.

CACTACEAE — CACTUS FAMILY

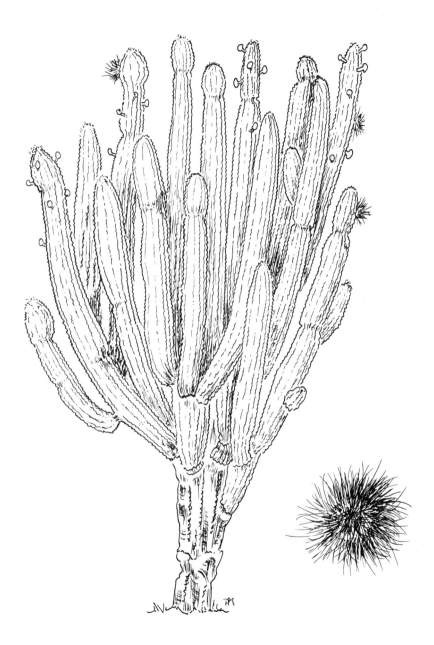

CACTACEAE — CACTUS FAMILY

Pereskia lycnidiflora DC.
Árbol del Matrimonia

The spiny trunk of a large tree with mostly horizontal branches and fleshy leaves will make one look twice to determine that *Pereskia* really is a cactus. The spines in various sizes and numbers are in areoles. Those clothing the trunk may reach 10 cm (4 in) in length and are black; those on the branches usually are smaller and frequently subtend a leaf or flower branch. Distinct ovate to obovate, deciduous, fleshy leaves on short petioles help to establish *Pereskia* as one of the primitive members of the Cactus Family.

Individual flowers adorning the tips of young branches appear on sunny days in June and July. Each flower has numerous, orange, fringed, perianth segments surrounding many pale yellow stamens. These all arise from the top of a receptacle clothed in bracts. From the receptacle, a globose to turbinate aromatic fruit develops, which at maturity turns greenish-yellow or red-orange.

Pereskia has been so widely cultivated it may be found in many areas of Mexico. Evidence indicates it to be a native of the deciduous shrub vegetation of Oaxaca.

Rhipsalis baccifera (J. Miller) W.T. Stearn
Tatache or Mistletoe Cactus

Rhipsalis, its multiple branches creating a mass of stems, from a distance resembles a large swarm of bees hanging from tree trunks and large branches in the evergreen tropical forests. The pencil-like stems lengthen by producing two or three new joints at the apex. There are no noticeable leaves, but minute scales are present. Scattered, widely-spaced areoles, with tufts of six to nine white, bristlelike hairs, dot the stem. From these areoles develop small flowers with relatively few perianth parts, which grade from greenish-white to white. Within are three series of stamens with about five stamens each. At maturity the flowers open widely like small stars. Below the perianth and bearing one or two small green scales is the ovary, which develops into a small, whitish, somewhat translucent, globose fruit (*baccifera* = berry bearing).

CACTACEAE—CACTUS FAMILY

Pereskia lycnidiflora
TREE to 10 m (32 ft) tall; trunk well defined; branches horizontal. LEAVES 4–7 cm (1.5–2.75 in) long, 5 cm (2 in) wide, ovate to obovate, fleshy. FLOWER 6 cm (2.5 in) in diameter, orange. FRUIT 3–4 cm (1.25–1.5 in) in diameter, globose.

CACTACEAE—CACTUS FAMILY

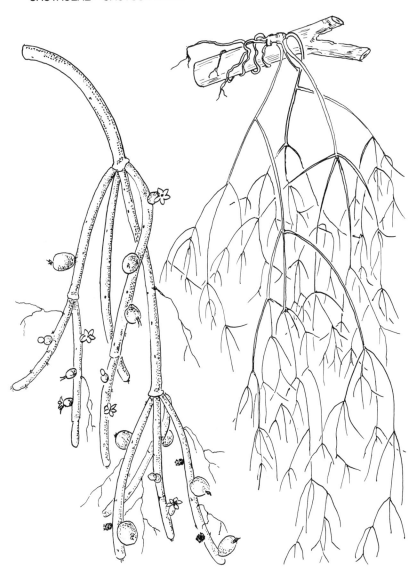

Rhipsalis baccifera
EPIPHYTE, pendant, 1+ m (3+ ft) long; joints 10–20 cm (4–8 in) long, 3–4 mm in diameter. FLOWER 5 mm in diameter. FRUIT 5 mm in diameter, globose.

Rhipsalis is the only genus of Cactaceae that occurs natively outside of the Americas; *R. baccifera* is found in Sri Lanka and in Africa, where a few other species grow. All of these can be traced to American species, leading to the inference that the species were widely distributed by migratory birds.

In Mexico, *R. baccifera* has been reported in the tropical forest of the east coast from Tamaulipas to Campeche, including Hidalgo, San Luis Potosí, and Puebla. It also grows in the western state of Jalisco.

STENOCEREUS SPECIES

Stenocereus is a genus of columnar cacti without a common structural outline to permit ready identification. It forms a natural botanic group because of the morphology of the flower and fruit, which seldom are seen by the traveler. Three representative species have been selected for inclusion here on the basis of distinctive characteristics: *S. marginatus*, for its ubiquitous use in living fences; *S. thurberi*, for its trunkless shrublike form, and *S. weberi*, for its immense candelabra form.

Stenocereus marginatus (DC.) Berg. and Buxbaum
Organo

The straight, columnar, usually unbranched "organo" often is planted to form an impenetrable fence around homes, corrals, and fields. Four to seven broad ribs bear large, closely spaced and often confluent, gray, felty areoles, in which are one or two central spines surrounded by seven or eight short radial ones. Flowers appear at random along the ribs from the apex of the stem downward. The cylindrical floral tube is clothed with scales subtending tufts of hairs and small spines, and ends in perianth segments that grade from greenish-white to rose. Immature fruits are covered with small spines and hairs, which fall off as development continues. Both the flowers and globose, yellowish-red fruits are eaten.

"Organo" is commonly cultivated, resulting in a wide distribution, but it is native to the high Central Plateau from Guanajuato, Queretaro, and Hidalgo south to Puebla, Oaxaca, and Guerrero.

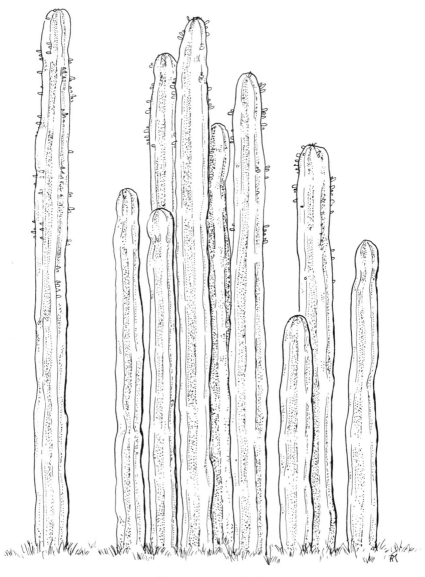

Stenocereus marginatus
COLUMNAR, 3–5 m (10–16 ft) tall, unbranched; ribs 4–7. SPINES short, 7 radial 2–4 mm long, 1–2 central 1–1.5 cm (.5–.75 in) long. FLOWER 3–5 cm (1.25–2 in) long. FRUIT 4 cm (1.5 in) in diameter, globose.

Stenocereus thurberi (Engelm.) Buxbaum
Pitaya Dulce or Organ Pipe Cactus

"Organ pipe cactus" is one of the three large cacti native to the western United States. Organ Pipe Cactus National Monument has been established in southwestern Arizona to protect this Mexican species in the northern limit of its distribution.

Stenocereus thurberi has many erect branches arising from ground level or from a very short trunk. Sixteen to nineteen low ribs at narrow intervals on each branch are topped with large, brown, felty areoles 1 to 3 cm apart. Seven to nine relatively short radial spines develop in each areole, surrounding one to three central spines. Flowers bloom on the upper parts of the branches from March to August. Blossoms open just before sunset and close during midmorning the next day, thereby taking advantage of both night pollinators, such as nectar-seeking bats, and day pollinators. The flowers have numerous perianth parts, ranging from the green scales that cover the bud, through reddish outer segments, to the purplish inner parts, the tips of which roll outward as the flower opens. Numerous white stamens line the inside of the floral throat. Globose edible fruits develop after pollination and at first are covered with spines; as the fruits mature the spines fall, leaving the fruit unprotected. Color varies from olive-green to reddish, with the crimson interior flesh surrounding many black seeds. The fruit is eagerly sought for its sweet flavor, and often is made into marmalade.

Restricted in distribution, "pitaya dulce" occurs in Sinaloa, Sonora, and Baja California.

Stenocereus weberi (Coulter) Buxbaum
Candelabro

A representative of the candelabra form of massive cactus, this *Stenocereus* species may produce a hundred or more vertical branches above a well developed trunk. Ten ribs form on each of the branches. From the widely spaced areoles, six to twelve stout, short, gray, radial spines surround the longer (10-cm [4-in]), single central spine. Large, white flowers open during the daytime to expose numerous stamens on short filaments. Edible, globose fruits with red-purple pulp and covered

CACTACEAE — CACTUS FAMILY

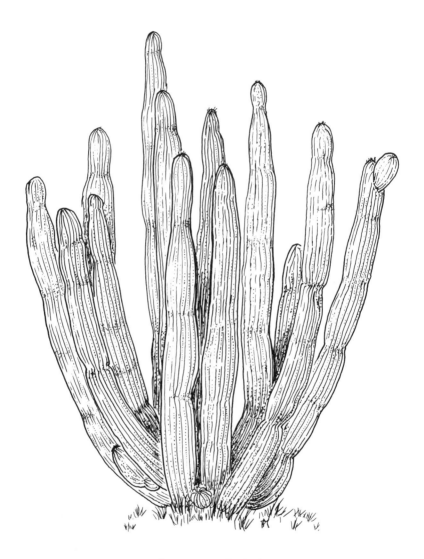

Stenocereus thurberi
COLUMNAR, many branches, 1–7 m (3–23 ft) tall, no trunk, ribs 16–19. SPINES 7–9 radial 1 cm (.5 in) long; 1–3 central 2–5 cm (.75–2 in) long. FLOWER 6–7.5 cm (2.5–3 in) long, night-blooming. FRUIT 2.5–7.5 cm (1–3 in) in diameter, globose.

CACTACEAE — CACTUS FAMILY

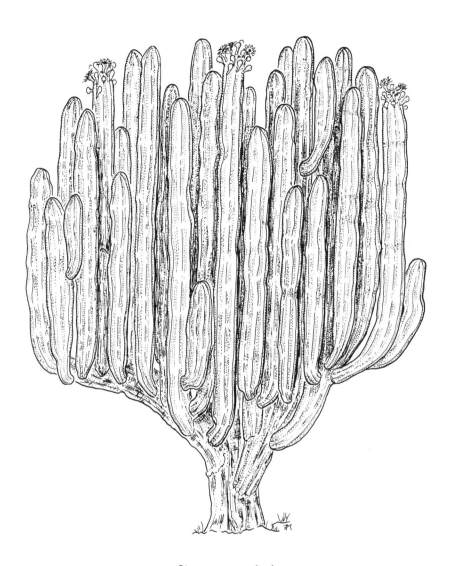

Stenocereus weberi
COLUMNAR, 10 m (32 ft) tall, trunk 2 m (6.5 ft) tall; branches erect, ribs 10. FLOWER 8–10 cm (3–4 in) long, white. FRUIT 6–7 cm (2.5–2.75 in) in diameter, spiny.

with abundant, long, yellow spines resemble those of *Pachycereus pectin-aboriginum.* Flowers and seeds of this gigantic cactus are used as forage. Because of its tremendous weight, the wood of its branches is sufficiently strong and durable to be used to roof the homes of local farmers. Stands of this stately cactus dominate the landscape in Puebla, Oaxaca, and Guerrero.

Caricaceae—Papaya Family

Carica papaya L.
Papaya

Palmlike in appearance, with an unbranched main trunk crowned with large leaves, this quick-growing tree yields a popular fruit. Palmately, deeply lobed leaves—each of its seven lobes again pinnately lobed—are dark green above, glaucous below, and borne on long hollow petioles. The trunk of the tree, with its soft fleshy tissue surrounding a central cavity, is covered with a pale green skinlike bark. The unisexual flowers may be both on the same tree (monoecious) or on separate trees (dioecious). Male flowers, sessile in long slender axillary racemes, have five united petals forming a yellow, funnelshaped corolla with lobes about half as long as the tube. Ten stamens are attached to the inside of the corolla tube. The female flowers, solitary or in few-flowered clusters, develop along the stem above the scar of the fallen leaves. Each flower has five yellow, separate, twisted petals.

"Papaya" fruit resembles a melon, being either elongated or globose with a bright orange or yellow, smooth skin. The thick orange interior flesh is sweet and encloses a central cavity containing many small, round, dark green or brown seeds embedded in a mucilaginous mass. Latex present throughout all the tissues of the plant contains papain, an enzyme similar to the digestive pepsin in an animal's stomach. "Papaya's" place of origin is unknown, but it now is grown throughout the tropics in home gardens, rather than in large groves. It often is seen along roadsides or near garbage dumps.

CARICACEAE—PAPAYA FAMILY

Carica papaya
TREE to 10 m (32 ft) tall, unbranched; white latex. LEAVES 20–75 cm (8–30 in) wide, deeply palmately lobed, glaucous beneath. FLOWER male 2–3 cm (.75–1.25 in) long; female 4–5 cm (1.5–2 in) long. FRUIT 5–60 cm (2–24 in) long, berry, elongate or globular.

Cochlospermaceae—Shell Seed Family

Cochlospermum vitifolium (Willd.) Kunth
Rosa Amarillo, Chimi, or Shell Seed

"Rosa amarillo" (yellow rose) aptly describes the showy, brilliant yellow flowers that bloom in the winter months, or dry season, on a leafless shrub or small tree. Each flower has dark brown sepals and five large, lemon-yellow, obovate, deeply notched petals, spreading like those of a wild rose; numerous incurving orange stamens form a colorful center. In Central America these stamens are used to adulterate or replace saffron. Racemes or axillary panicles of the flowers often create a brilliant mass of color.

The globose, dark brown fruit, covered with soft gray hairs, is five-parted and contains shell-like seeds. Each seed is embedded in soft, white, hairy floss, which is used regionally as cotton. The common name "shell seed" translates the Greek of the generic name *Cochlospermum* (*kochlion* = shell, *sperma* = seed). The large attractive leaves, deeply cut into five lobes with serrated edges, resemble grape leaves; *vitifolium* in Latin means "grapevinelike leaf."

Cochlospermum can be found from southern Sonora south to Oaxaca, Veracruz, and Yucatán, and southward through Central America to northern South America.

Combretaceae—Combretum Family

Combretum fruticosum (Loefl.) Stuntz
Peineta

Dense spikes of numerous, showy, red or red-orange flowers make *Combretum fruticosum* a prominent member of the spring flora. The flowers are closely arranged along one side of the axis; although the four or five red petals are minute, eight or ten long, showy, crimson stamens make the inflorescence conspicuous, resembling an ornamental comb (*peineta*). The small calyx, four-winged fruits, and undersides of the leaves are densely covered with scales.

COCHLOSPERMACEAE—SHELL SEED FAMILY

Cochlospermum vitifolium

SHRUB or tree to 25 m (82 ft) tall. LEAVES 10–30 cm (4–12 in) wide, 5-lobed. FLOWER 8–12 cm (3–5 in) in diameter, yellow. FRUIT capsule 8 cm (3 in) in diameter, 5-parted.

COMBRETACEAE—COMBRETUM FAMILY

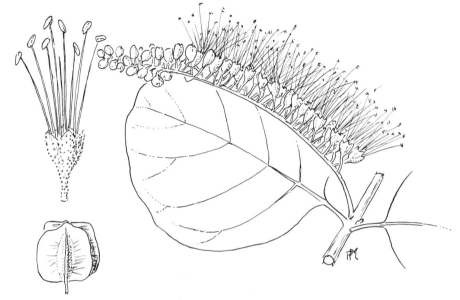

Combretum fruticosum

VINE woody, unarmed. LEAVES 5–15 cm (2–6 in) long, opposite, oval to elliptic-oblong. FLOWER red, minute, in brushlike spikes. FRUIT 2 cm (.75 in) long, 4 wings.

Striking, sweet-scented flowers, full of nectar, attract insects and hummingbirds. The stem, when cut, is said to yield a considerable amount of water and is flexible enough to be woven into coarse baskets or substituted for rope. *Combretum* grows from Sinaloa to Chiapas and Veracruz, and south into Central America.

Conocarpus erecta L.
Botoncillo or Buttonwood

A plant of the mangrove swamps, *Conocarpus* varies from a prostrate shrub to an erect tree. It has dark brown fissured bark with thin scales. The leathery leaves, either obovate or elliptic, are on short petioles, which bear two prominent glands at the base of the blade. Many minute green flowers are clustered in dense, conelike heads, which are arranged in terminal or lateral inflorescences. Petals are absent, but five to ten stamens stand prominently erect from the green calyx.

Conocarpus erecta

SHRUB or tree 20 m (65 ft) tall, in coastal swamps. LEAVES 2–10 cm (.75–4 in) long, alternate, leathery. FLOWER inconspicuous, green. FRUIT conelike.

The small heads turn from green to greenish-purple as the flat, winged scalelike drupes mature. The mature heads resemble buttons.

"Buttonwood" grows in the swampy areas of both coasts, from Tamaulipas and southern Baja California south through Central America. It is one of the four common shrubs or trees of the coastal mangrove swamps; see also *Avicennia*, *Laguncularia*, and *Rhizophora*.

COMBRETACEAE—COMBRETUM FAMILY

The wood is satisfactory for many uses: boat-building, fences, wood-turning, cross ties, and cabinet work. Because it burns slowly, the wood makes good fuel and charcoal. The leaves and bark are useful in tanning and the bark is used medicinally. *Conocarpus erecta* and a variety with silvery, silky hairs may serve as cultivated ornamentals in areas away from the seashore.

Laguncularia racemosa (L.) Gaertn.
Mangle Blanco or White Mangrove

This evergreen tree, one of the four common members of the coastal mangrove swamps, is characterized by opposite, leathery, elliptical leaves on short stiff petioles bearing a pair of glandular dots at the apex. Terminal clusters of fragrant flowers usually are branched and spreading. The small, whitish, bell-shaped flowers are stalkless, with a minute five-lobed calyx and five rounded white petals. The gray-green or brownish fruit is slightly fleshy, small, pear-shaped, and at maturity has ten ridges. It contains but one seed, which may enlarge and germinate while the fruit is still attached to the tree or in the water where it drops and floats.

"White mangrove" in the shoreline swamps of both coasts usually is associated with *Rhizophora*, *Avicennia*, and *Conocarpus*. It grows with its stilted roots in water, but is unable to survive in water as deep as that in which *Rhizophora* grows. It is found from Sonora and Baja California south along the west coast, and from Tamaulipas south to Central America along the east coast.

Compositae—Sunflower Family

Ageratum corymbosum Zuccagni
Cielitos

This showy plant, with its numerous colorful flower heads, may have soft herbaceous stems arising from a ground-level, woody base (caudex), or it may grow as a shrub. The stems, red toward the upper portion, bear opposite, variable leaves ranging from narrowly lanceolate

COMPOSITAE – SUNFLOWER FAMILY

Laguncularia racemosa

TREE 20 m (65 ft) tall; bark gray-brown, rough, fissured. LEAVES 4–10 cm (1.5–4 in) long, opposite, elliptic, leathery. FLOWER small, white. FRUIT 15 mm long, sepals persistent.

to broadly ovate, with entire, crenate, dentate, or lobed margins. The variability of the leaves has led to the recognition of five taxonomic forms based on leaf shape. Small heads of regular flowers clustered into terminal corymbiform inflorescences (*corymbosum* = flat-topped cluster) bloom from June to January. Each hemispheric head has two series of bracts enclosing several small funnelform flowers. The corolla

COMPOSITAE — SUNFLOWER FAMILY

Ageratum corymbosum

HERB perennial, or shrub to 2 m (6.5 ft) tall. LEAVES variable 3.5–10 cm (1.5–4 in) long, 1.5–4.5 cm (.75–2 in) wide, opposite. FLOWER 2–3.75 mm long, blue; inflorescence corymbose cluster of heads. FRUIT 1.3–3 mm long, achene, 5-angled, dark brown to black.

has a green to whitish tube expanding into a white pilose throat tipped with five spreading blue, lavender, bluish-gray, or white lobes. At maturity, five-angled dark brown or black achenes are formed.

This feathery-flowered perennial herb or shrub is a favorite for cultivation in gardens. The most common native species of *Ageratum* in Mexico, it flourishes throughout the country, except in the Yucatán Peninsula and Baja California.

Encelia farinosa A. Gray
Hierba Ceniza or Brittle Bush

"Brittle bush" is a rounded, low-growing shrub with gray or silvery-white leaves crowded near the ends of branches. Long naked stems elevate the numerous flowering heads above the lanceolate or broadly ovate, felty leaves. (The specific name *farinosa* means powdery or mealy.) Each head has several bright yellow ray flowers surrounding a yellow cluster of disk flowers. In one variation of this species, the yellow center is replaced by purple disk flowers. From the ends of mature stems a brownish resin may exude, which served the early natives as a glue and as an incense for religious purposes.

Blooming in winter and early spring, this showy shrub is very conspicuous on rocky hillsides in the desert areas of Baja California, Sonora, and Sinaloa.

Flourensia cernua DC.
Hojasé or Tar Bush

Flourensia, a glutinous shrub with a tarlike odor, hence the common name "tar bush," bears small heads of yellow flowers. Although not showy, it is conspicuous by its abundance on the Chihuahuan Desert. Arising from leaf axils and terminating the short leafy branches are pendant, bell-shaped heads composed of yellow disk flowers surrounded by involucral bracts (phyllaries) in three rows. Achenelike fruits, clothed in silky hairs, follow the flowers upon pollination. The small alternate leaves with entire margins are ovate to ovate-lanceolate in outline.

Growing from Sonora to Nuevo Leon and south to Durango, Zacatecas, and San Luis Potosí, this species flowers from September to December.

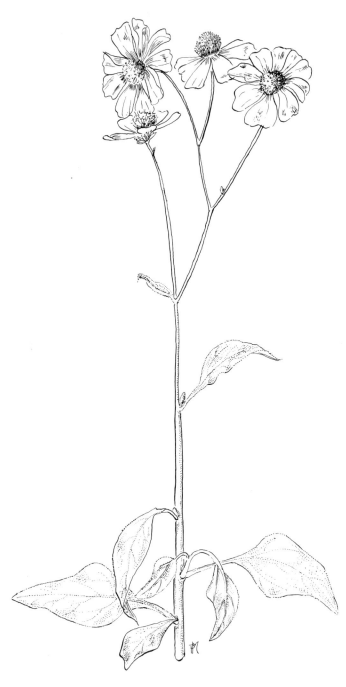

COMPOSITAE – SUNFLOWER FAMILY

Flourensia cernua

SHRUB 1–2 m (3–6.5 ft) tall. LEAVES 1.5–2.5 cm (.75–1 in) long, 8–15 mm wide, alternate, glutinous. FLOWER head 8–10 mm long; flowers all disk, yellow; rays lacking. FRUIT 6 mm long, 2 mm wide, achene.

Encelia farinosa

◀ SHRUB 2 m (6.5 ft) tall. LEAVES 1.5–2.5 cm (.75–1 in) wide, gray, felty. FLOWER head 5 cm (2 in) in diameter, yellow.

Montanoa tomentosa Cervantes
Zoapatli

Montanoa, a common roadside shrub, has numerous, small, sparsely to densely tomentose heads of fragrant flowers in broad, compound, terminal inflorescences. Small cream to white ray flowers contrast with the light to dark yellow flowers of the central disk. Usually only one brown-black achene per head matures. The slender, petioled leaves, opposite in arrangement, are triangular-ovate in shape, serrate or irregularly toothed on the margin, and may be variously lobed or entire.

This shrub is found from Sonora southward to Oaxaca and Chiapas on the west, and from San Luis Potosí southward through Hidalgo, Veracruz, and Puebla to Oaxaca on the east. The genus was named by Cervantes in 1825 to honor Don Luis Montaña, a physician and naturalist from Puebla.

Pluchea odorata (L.) Cass.
Santa Maria

A strongly aromatic shrub with large flat-topped clusters of small purplish flowers, *Pluchea odorata* can be seen blossoming at almost any season of the year. Numerous flowers, clustered into small composite heads, are surrounded by five or six series of phyllaries. Dirty white, silky hairs crown each ovary. Ovate-oblong to elliptic leaves, with bases attenuated along the stout petiole, usually have entire or distantly toothed margins. The stem, inflorescence, and lower surface of the leaf are densely covered with curly brown hairs.

As this is an aromatic plant with the specific epithet *odorata,* history is bound to ascribe to it some medicinal value; the literature is full of reports of its use to cure everything from the bites of venomous animals to stomach ailments. Widely distributed in Mexico and usually growing in canyons, marshy areas, and along arroyo banks, *Pluchea* is found from Baja California to Tamaulipas and south to Veracruz, Yucatán, and Chiapas.

COMPOSITAE – SUNFLOWER FAMILY

Montanoa tomentosa

SHRUB 3–8 m (10–25 ft) tall. LEAVES highly variable; petioles 1.5–4 cm (.75–1.5 in) long; blade 3–20 cm (1.25–8 in) long, 1.5–15 cm (.75–6 in) wide. FLOWER head 3–8 mm wide; rays 3–5 mm long, 0–6 per head; disk flowers 7–16 per head; panicles 30–40 cm (12–16 in) wide. FRUIT achene brown-black, 2.5–3.5 mm long.

Pluchea odorata

SHRUB 1−2.5 m (3−8 ft) tall, hairy. LEAVES 7−15 cm (3−6 in) long, 2.5−6 cm (1−2.5 in) wide, entire. FLOWER small, purple; heads 7 mm high; inflorescence 6−15 cm (2.5−6 in) broad, corymbose. FRUIT less than 1 mm long.

Senecio praecox (Cav.) DC.

Candelero

Senecio praecox flowers while the plant is leafless, producing umbel-like clusters of yellow heads on the ends of stubby branches. Each head is composed of three to five ray flowers around the outside of a center containing six to eighteen disk flowers. The plants are shrubs

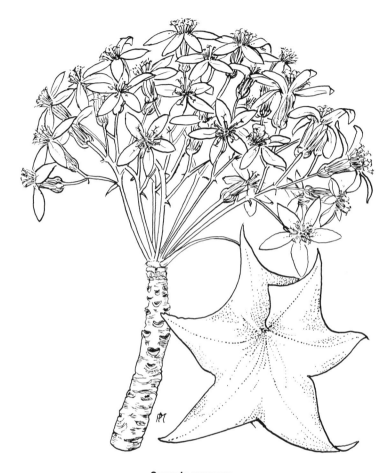

Senecio praecox

SHRUB or small tree 2−6 m (6.5−20 ft) tall. LEAF blade 10−15 cm (4−6 in) long, 3−5 cm (1.25−2 in) wide, ovate to suborbicular. FLOWER yellow; rays 3−5, 7−13 mm long; disk flowers 6−18, 9−12 mm long; inflorescence terminal umbellate, 5−10 cm (2−4 in) high, 10−20 cm (4−8 in) broad. FRUIT 4−6 mm long, achene.

or small trees, with thick, heavy-looking branches terminating in an inflorescence of flowers; later a cluster of ovate to almost round leaves develops, each with five to seven sharply pointed lobes. A section of the rubbery branch, if split lengthwise, shows the central, pithy portion to be divided into segments or chambers. At maturity the small brown achenes are topped by a tuft of white hairs (pappus).

This plant grows on cliff faces, lava fields, or steep, rocky hillsides from Nayarit, Jalisco, and Zacatecas south to Oaxaca. It may be found in flower from February to May.

Senecio salignus DC.
Jarilla

Senecio salignus, a weak-stemmed, spreading, often drooping shrub, produces masses of showy yellow composite flowers from January to July. The profusion of terminal inflorescences forms leafy panicles with many flower heads on the ends of short branches. Each head consists of four to six rays circling the 15 to 20 central disk flowers. Abundant willowlike (*salignus* = willowlike), linear to elliptic-linear leaves taper at both ends. They are glabrous, clear green, with entire margins. Small, brown, ribbed achenes develop from both the ray and disk flowers.

This species grows from the southwestern United States south to Chiapas and into Central America. It is common throughout the Central Plateau of Mexico in arid brushlands, oak grasslands, and often is weedy along roadsides and clearings.

Tithonia fruticosum Canby & Rose
Mirasol or Tree Sunflower

Large yellow composite flowers, resembling sunflowers growing profusely on a treelike shrub, mark the "tree sunflower." The conspicuous heads, supported by swollen peduncles, have showy yellow rays surrounding a large center of many yellow disk flowers. Each of the individual disk flowers is subtended by a bract tipped with dark brown, about twice as long as the mature achene (fruit), giving the older composite head the appearance of having a brown center. Surrounding the base of the complete head are green, broadly ovate to obovate bracts (phyllaries) overlapping like shingles and arranged in four rows or series. Felty, dark green leaves change from an opposite arrangement in the lower branches to alternate near the top of the plant. Growing on ex-

COMPOSITAE—SUNFLOWER FAMILY

Senecio salignus

SHRUB 1−2.5 m (3−8 ft) tall. LEAVES 3−13 cm (1.25−5 in) long, 2−10 mm broad, linear to elliptic-linear. FLOWER yellow; rays 4−6, 5−6 mm long; disk flowers 15−20, 5−6 mm long; inflorescence 10−20 cm (4−8 in) high, multicapitate terminal panicle. FRUIT 3.3−3.8 mm long, 1 mm thick, ribbed, achene.

panded, winglike petioles, they are ovate to lanceolate, with attenuate, often curved tips and finely crenate margins.

These showy sunflowerlike shrubs grow along water courses and in shaded canyons of Sonora, Chihuahua, Sinaloa, and Durango. Many species of *Tithonia* range over Mexico, varying in size and in color (yellow, orange, red); they can be distinguished readily from other composites by the abundance of soft, velvety leaves and the swollen, hollow peduncle supporting each head.

COMPOSITAE—SUNFLOWER FAMILY

Tithonia fruticosum

SHRUB 3–4 m (10–13 ft) tall. LEAVES 6–30 cm (2.5–12 in) long, 3–5 cm (1.25–2 in) broad. FLOWER yellow; rays 14–20, 2–3.5 cm (.75–1.5 in) long; disk 2–2.5 cm (.75–1 in) high, 2.5–4.5 cm (1–2 in) wide; head 7–9.5 cm (2.75–4 in) wide. FRUIT 4–5 mm long, achene.

Convolvulaceae—Morning-glory Family

Exogonium bracteatum (Cav.) Choisy
Bejuco Blanco

Thickly flowered, colorful racemes space themselves at frequent intervals along the usually leafless stems of this woody vine. Pink or purple bracts overlap like shingles; as in *Bougainvillea*, these bracts are the showy part of the *Exogonium* inflorescence. Even the specific name emphasizes these floral parts; *bracteatum* means bearing bracts. The red, narrow, tubular corolla of five united petals protrudes from the bracts. The exserted stamens and stigma add to the beauty of this flower, which blooms from December to May.

This vigorous vine may spread over the ground or climb in adjacent trees and shrubs. It has a large, watery, sweet root, which reportedly is eaten either raw or cooked. Heart-shaped, glabrous, light green leaves have long attenuated tips.

Exogonium bracteatum occurs from Baja California to Chihuahua, south to Veracruz and Oaxaca, but more commonly in western Mexico.

Ipomoea arborescens (Humb. & Bonpl.) D. Don
Palo Blanco or Tree Morning-glory

Late autumn or winter travelers in Mexico will be startled to see a starkly leafless, white-barked tree with large white morning-glory flowers. The conspicuous blooms appear in few- to many-flowered panicles, usually only one flower opening in the cluster at a time. A long greenish or cream tube supports the spreading, white, ruffled corolla; in its darkish throat the stamens are enclosed.

The globose or elongated fruiting capsule is narrowly ovate and contains four dark reddish-brown seeds. Each seed bears a fringe of white or yellowish silky hairs on the upper two-thirds of the exterior angles. Leaves are ovate, cordate at base, densely pubescent beneath, and have prominent veins.

Ipomoea arborescens, one of the many species of this genus in Mexico, may be found from Sonora and Chihuahua south to Veracruz and Oaxaca.

CONVOLVULACEAE — MORNING-GLORY FAMILY

Exogonium bracteatum

VINE glabrous, deciduous. LEAVES 6–9 cm (2.5–3.5 in) long, ovate-cordate. FLOWER 3–5 cm (1.25–2 in) long, red; bracts 2–3.5 cm (.75–1.5 in) long, showy, prominently veined.

CONVOLVULACEAE — MORNING-GLORY FAMILY

Ipomoea arborescens
TREE to 12 m (40 ft) tall, deciduous. LEAVES 8–20 cm (3–8 in) long, ovate. FLOWER 6–12 cm (2.5–5 in) long, showy, white. FRUIT 15–22 mm long.

Ipomoea pes-caprae (L.) R. Br.
Pata de Vaca or Hierba de la Raya

A profusion of bright pink to violet flowers of the morning-glory type cover the sprawling, ground-hugging *Ipomoea pes-caprae*. Large erect flowers on stout pedicels, borne singly or in several-flowered clusters, are broadly funnelform; widely spreading corollas have ruffled edges. From December to April the profusion of bright blooms makes

CONVOLVULACEAE — MORNING-GLORY FAMILY

Ipomoea pes-caprae
VINE, trailing, prostrate; stems 1–1.5 cm (.5–.75 in) in diameter. LEAVES 6–10 cm (2.5–4 in) long, orbicular, succulent. FLOWER 2.5–5 cm (1–2 in) long, limb 5–8 cm (2–3.5 in) broad, pink. FRUIT 1.5–2 cm (.5–.75 in) long, capsule, ovoid; seeds densely pubescent.

a colorful display as the vine sprawls, with numerous branches, up to 20 m (65 ft) along sandy upper beaches and coastal dunes. The stems are coarse, ropelike, and strong enough to withstand the turbulent winds that lash the shore. Each node has the potential not only for producing a flower, but also for root development, which helps to anchor the long stems.

Round, leathery, glabrous leaves, long-petiolate and notched at the apex, fold at the midvein to give the appearance of two connected leaves. When open flat, they resemble the hoof print of a cow, hence the common name "pata de vaca." A milky latex is present in the leaves and stems. Round, brown, nearly woody capsules on thickened pedicels contain four densely pubescent seeds.

Ipomoea pes-caprae is a characteristic and widely distributed plant of tropical seashores, and as such is found not only on the sandy beaches and dunes of Mexico, but around the world.

Cyatheaceae—Tree Fern Family

Nephelea mexicana (Schl. and Cham.) Tryon
Cola de Mono

Although it appears at first glance to be a palm tree, *Nephelea*, a fern, is readily distinguished from the palms by its leaf structure and growth. Large, feathery leaves have bipinnate blades with pinnae also dissected. Both the stem and petiole are protected by blackish spines. Young leaves arising from the crown uncurl as they grow, resembling the curled tail of a monkey, giving rise to the common name "cola de mono" (tail of the monkey). Ferns have neither flowers nor seeds but reproduce by spores, which are produced in great abundance in little "fruit dot" areas on the lower surfaces of the fertile pinnae.

This fern grows in the tropical deciduous forests and has been reported from San Luis Potosí, Hidalgo, and Veracruz south through Puebla to Oaxaca and Chiapas.

CYATHEACEAE—TREE FERN FAMILY

Nephelea mexicana

TREE 10 m (32 ft) tall. LEAF petiole 1–2 m (3–6.5 ft) long, spiny; blade 3–4 m (10–13 ft) long, bipinnate pinnatifid.

Ericaceae—Heath Family

Arbutus xalapensis H.B.K.
Madroño

Smooth brown bark peeling in large, thin sheets is characteristic of this species of *Arbutus*, as are the brown or white glandular hairs that cover the reddish, tomentose branchlets. Hairs also appear on the undersides of the leathery, broad-oblong, ovate, or ovate-lanceolate leaves, which remain glabrous above. Usually evergreen, "madroño" may lose its leaves after a prolonged dry season. Tomentose, glandular hairs also cover the paniculate inflorescences, which are densely flowered with delicate, white or pink, small, urnshaped blooms. The short tube of the calyx is formed by five deeply cut lobes. The squat, tubular corolla of five united petals with lobes reflexed resembles a small fat lady with a ruffled collar and a tight belt around her middle. The flower is followed by a red, globose berry.

The common name "madroño" is applied to most of the ten species of *Arbutus*, wherever found. *A. xalapensis* grows from Sinaloa, Chihuahua, and Nuevo Leon south through Veracruz and Oaxaca to Guatemala.

Euphorbiaceae—Spurge Family

Euphorbia antisyphilitica Zucc.
Candelilla

Euphorbia antisyphilitica, a queer, low, rhizomatous plant with milky latex, forms large clusters of round stems with the diameter of an ordinary lead pencil. These gray-green stems, either branched or unbranched, are coated with a thick waxy layer. Small, linear, awl-shaped (subulate) leaves form simultaneously with new growth, but these soon disappear, leaving the plant essentially leafless. During the flowering season in June or July, one to three small cuplike structures containing several staminate and one pistillate flower develop at purplish nodes along the stem. Each cup is rimmed by five glands, four of equal

Arbutus xalapensis
TREE or shrub 4–15 m (13–50 ft) tall. LEAVES 8–14 cm (3–3.5 in) long, oblong, ovate, or ovate-lanceolate, leathery. FLOWER in panicles 3–8 cm (1.25–3 in) long; calyx 4–6 mm broad; corolla 6–8 mm long, urnshaped, pink or white. FRUIT 8–12 mm in diameter, red, globose.

size and another half their size. Small, white, petal-like appendages, 1.5 to 2.5 mm long, decorate each gland.

The abundant wax on the stems makes extraction for commercial purposes worthwhile, to the detriment and destruction of natural populations. Candles, shoe polish, floor polish, lubricants, waterproofing, and an insulating agent in electrical apparatus are among the uses for "candelilla" wax.

Primarily a plant of the Chihuahuan Desert, this strange *Euphorbia* grows from Chihuahua and Coahuila south to Hidalgo and Puebla.

EUPHORBIACEAE—SPURGE FAMILY

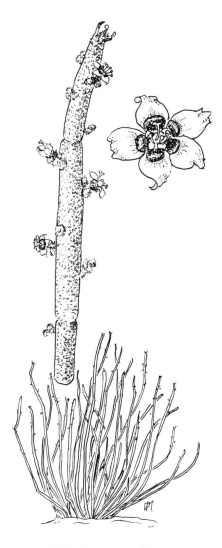

Euphorbia antisyphilitica

STEM to 1 m (3 ft) tall, 3–4 mm in diameter, numerous, cylindrical, pencil-like; milky latex. LEAVES 2–3 mm long, ephemeral, awl-shaped. FLOWER small, unisexual, clustered in campanulate cups 1–1.3 mm long. FRUIT 3.5–4 mm long, green or purplish.

EUPHORBIACEAE — SPURGE FAMILY

Euphorbia pulcherrima **Willd.**
Flor de Pascua or Poinsettia

One of the favorite plants of Mexico is the native Christmas poinsettia, *Euphorbia pulcherrima*, a plant widely cultivated for its beauty. Few people realize that the showy red, pink, white, or pale yellow parts are not flowers, but floral leaves (bracts). Hidden among these bracts are inconspicuous unisexual flowers clustered in small yellow cups. Each involucral cup has one large, conspicuous gland on the side. Several male flowers, with single stamens, and one female flower protrude from the mouth of the chalice. The ovary of the pistillate flower is three-lobed with six stigmas. All parts of the plant contain a milky latex that can be irritating to susceptible individuals.

The common name "flor de pascua" (Christmas flower) refers to the abundance of showy red bracts at the Christmas season; the plant is naturally conditioned to bloom during shortened daylight hours. The other common name, "poinsettia," which at one time was the generic name, was given to honor Joel R. Poinsett, United States Minister to Mexico, who "discovered" the plant in 1828 and introduced it to cultivation in the north. "Poinsettia" grows natively from Jalisco to Oaxaca and Veracruz. It is widely cultivated out of doors in frost-free areas, and in containers in colder regions.

Hura polyandra **Baill.**
Jabillo or Sand Box Tree

When in fruit, curious wheel-like or melon-shaped capsules with numerous sections delineated like a peeled orange call attention to *Hura polyandra*. At maturity the dry capsule explodes with a loud report and tremendous force, scattering the segmented sections and their large, flat, brown seeds to great distances. Also unusual on this deciduous tree are the flowers, which, typical of this family, are of two types, male and female, both borne on the same tree. Neither has petals. The small, staminate (male) flowers, clustered on an elongated fleshy spike, have numerous stamens. The solitary female flower is of most unusual shape, for the styles are fused and their stigmas radiate outward like the outstretched ribs of an umbrella. The ovary, with approximately 15

EUPHORBIACEAE – SPURGE FAMILY

Euphorbia pulcherrima

SHRUB 1–8 m (3–26 ft) tall; milky latex. LEAVES 12–20 cm (5–8 in) long, ovate; floral leaves red, pink, whitish, or pale yellow. FLOWER small, unisexual, clustered in yellow cups. FRUIT 3-lobed.

cells, produces the odd wheel-shaped fruit. The straight gray trunk of the tree frequently is covered with short, broadly conical, sharp spines. Large, alternate, simple, heart-shaped leaves on long petioles have toothed margins; the blade is almost as wide as it is long.

The milky latex present in *Hura* is caustic and may produce blisters on susceptible people. It also is poisonous, and is mixed with sand or

EUPHORBIACEAE—SPURGE FAMILY

Hura polyandra

TREE 15–30 m (50–100 ft) tall; crown spreading. LEAVES 11–20 cm (4.25–8 in) long, heart-shaped; petioles 8–15 cm (3–6 in) long. FLOWERS unisexual on same tree. FRUIT 8–10 cm (3–4 in) in diameter, wheel-shaped.

meal to be used as a fish poison. The seeds, which contain 50 percent oil, also are dangerously poisonous. In the Dominican Republic, children use the immature green fruits as wheels for their toys. The common name "sand box tree" arose from the former use of the dried immature fruits, from which the seeds had been removed, as vessels to hold fine sand for drying or blotting ink.

This interesting tree can be found from Sinaloa south to Chiapas, into Veracruz and Yucatán.

EUPHORBIACEAE – SPURGE FAMILY

Jatropha dioica Cervantes
Sangre de Drago or Rubber Stem

If an area looks as if someone had randomly stuck numerous pegs with fuzzy tops into the ground, it probably is a patch of *Jatropha dioica*, a peculiar perennial with horizontal rootstocks from which many thick leathery stems arise. The round fleshy stem is flexible (hence the common name "rubber stem") and unbranched except for a few short lateral spurs where clusters of spatulate or linear deciduous leaves grow. As the plant is unisexual (the species name is derived from the word *dioecious*), there are two separate types of inflorescences. The hairy staminate flowers, blooming in terminal or axillary clusters, have five white recurving petals and ten stamens. In contrast, the urnshaped pistillate flower with recurving lobes usually is borne singly on the lateral spurs. As a result of pollination, a globose fruit forms with an extended, curved remnant of the style at the apex and a keel along the sides. All parts of the plant are infused with a clear sap that turns blood red when it oxidizes, giving rise to the common name "sangre de drago" (blood of the dragon).

This unusual wandlike plant is widespread in desert areas of the Altiplano, from Chihuahua and Coahuila south to San Luis Potosí, and in the dry areas of Oaxaca.

Pedilanthus macrocarpus Benth.
Candelillo

Almost everything about *Pedilanthus macrocarpus* is unusual. The gray, cylindrical, jointed stems, arising erect in clumps from an underground woody root crown, are leafless and unbranched. When injured, the waxy, turgid stems exude an abundance of poisonous, milky latex. Small ephemeral leaves fall almost as soon as they are formed, leaving the naked stems unadorned until their flowering period (February to May and August to October), when one to five red flowers appear. The unisexual blooms are borne in clusters of several male and one female flower, giving the illusion of a perfect flower with stamens and umbrella-shaped pistil protruding from the toe of a bright red, slipper-shaped structure called a cyathium, hence the generic name *Pedilanthus* (Greek

Jatropha dioica
HERB to 1 m (3 ft) tall; perennial horizontal rootstock 1 m (3 ft) long; clear latex turns red. LEAVES 4–5 cm (1.5–2 in) long, spatulate, clustered on spur branches; deciduous. FLOWERS dioecious; staminate 1.5 mm long, white, clustered; pistillate single, urnshaped. FRUIT 1–1.2 mm thick; capsule, 1 or 2 cells.

Pedilanthus macrocarpus
SHRUB .5–1.5 m (1.5–6 ft) tall; stems fleshy 1–1.5 cm (.5–.75 in) in diameter; ▶ milky latex. LEAVES 3–10 mm long, spatulate to ovate, ephemeral. FLOWERS unisexual, clustered in red, slipper-shaped cyathium 10–23 mm long. FRUIT 7–15 mm long, 12–20 mm wide; seeds 6–9 mm long.

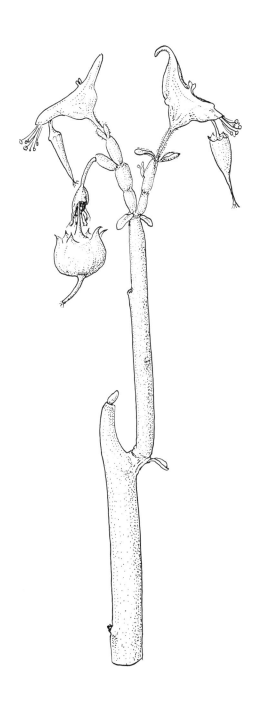

pedilon = slipper or sandal, *anthus* = flower). As the indehiscent, corky red fruit matures, the umbrellalike pistil expands into a globose structure topped with recurving points and containing three large globose seeds. A plant of the Sonoran desert, *Pedilanthus* grows throughout coastal Sonora south to northern Sinaloa, and abundantly on the peninsula of Baja California.

Ricinus communis L.
Palma Cristi, Ricino, or Castor Bean

Ricinus communis, a native of the Old World tropics, varies greatly in structural size, foliage coloration, and longevity, depending upon its habitat. Large palmately lobed leaves, coarsely toothed on the margins, may be green, dull red, purplish, or bronze; they are borne on strong petioles as long as, or exceeding the diameter of, the leaf blade. The hollow stems, petioles, and leaf margins often have stalked or sessile glands. Small unisexual flowers without petals form in terminal paniculate racemes. A five-parted calyx of the staminate flower subtends the numerous densely branched stamens. Pistillate flowers, generally located above the staminate, develop a slightly three-lobed ovary capped with three conspicuous, hairy, bright red stigmas. Large colorful clusters of fruiting capsules, with their reddish prickly exteriors, are as handsome as a bouquet of flowers. Each of the three chambers of the mature fruit contains one brown, mottled, smooth, ellipsoid seed, which is forcibly ejected when the dry capsule splits. It is this seed that contains a useful oil.

All parts of the plant, particularly the seeds, contain the highly poisonous phytotoxin ricin. The oil expressed from the seeds is prized for its lubricating properties, in addition to its well known use as the laxative castor oil. The name *Ricinus* also is the generic name of a small brown tick that infects sheep and dogs; the plant was named for the similarity in appearance of its seed and the tick. Widely distributed throughout Mexico as a cultivated plant for ornamental or economic purposes, *Ricinus* also can be seen along the roadsides, where it has become established as an escape.

EUPHORBIACEAE — SPURGE FAMILY

Ricinus communis

HERB, shrub, or tree 1–10 m (3–32 ft) tall; stems, branches hollow, brittle. LEAVES to 80 cm (30 in) in diameter, orbiculate, 7–10 palmate lobes. FLOWER small, monoecious; staminate mostly below pistillate. FRUIT 1.5–2.5 cm (.5–1 in) in diameter, prickly; seed 10–17 mm long.

Fagaceae—Oak Family

QUERCUS SPECIES

Quercus, the genus of oaks, is so large and complicated that identification of the species is difficult. Members of the genus may be trees or shrubs, evergreen or deciduous, with extremely variable, lobed or entire, usually leathery leaves. The wood is hard, and is suitable for furniture, for construction of all kinds, and for firewood. The genus can be recognized by its flowers of two kinds: small male flowers are clustered into ephemeral, elongated, pendant catkin inflorescences; female flowers are single or clustered, stalked or sessile, and at maturity form the familiar acorn, with its accompanying cup composed of many durable woody bracts. Neither male nor female flowers are showy, nor do they have petals.

From the large number of oaks in Mexico, six species have been selected for the distinctive shapes of their leaves, for their large or otherwise outstanding acorns, or for their broad distribution.

Quercus albocincta Trel.
Encino Negro

Quercus albocincta is a deciduous oak of northwestern Mexico. Twelve to eighteen large teeth, each tipped with a bristle almost 2 cm (1 in) long, make the elliptic-lanceolate leaves a conspicuous feature of this tree. An added distinction is the network of whitish veins in the leaves and a white, hard, cartilaginous margin, which prompted the specific name (*albo* = white; *cincta* = go around). Small, sweet, edible acorns require two years on the tree to mature. This is a tree of the western Sierra Madre, recorded in Sonora and Sinaloa.

Quercus candicans Née
Encino Blanco

Quercus candicans is a moderate to large tree with deciduous leaves that often turn red before falling. The leaves vary in outline from ovate to elliptic or obovate, with tips that may be rounded, acute, or short-

FAGACEAE—OAK FAMILY

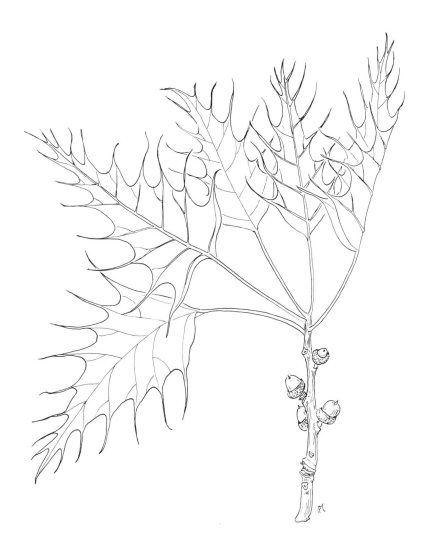

Quercus albocincta
TREE to 10 m (32 ft) tall. LEAVES 10–16 cm (4–6 in) long, 5–9 cm (2–3.5 in) wide, elliptic-lanceolate. FRUIT 8 mm long, 6 mm wide, sessile; cup ½ length of acorn.

FAGACEAE — OAK FAMILY

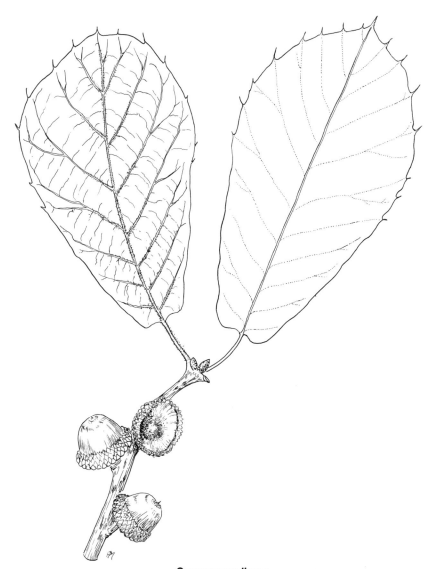

Quercus candicans
TREE to 20 m (65 ft) tall. LEAVES 11–25 cm (4.25–10 in) long, 4–15 cm (1.5–6 in) wide, ovate-elliptic to obovate. FRUIT 20–25 mm long, 16–18 mm in diameter; cup 20–25 mm in diameter, ⅓–½ length of acorn, sessile or very short-pedicelled.

acuminate. The upper half of the leaf is at times serrate and bristly. Hairs are abundant on the under surfaces of the leaves, making them appear creamy white (*candicans* = white hairy), while above they are a glabrous dark green. Ovoid acorns on very short pedicels are about twice as long as the cup in which they sit. "Encino blanco," one of the largest and most beautiful of oaks, grows from Sinaloa south to Guerrero and east to Veracruz.

Quercus conspersa Benth.
Encino Rojo

Quercus conspersa, a large oak of southern Mexico, has variable deciduous leaves that range from narrow or broadly lanceolate to ovate with long attenuate apices. The margins are smooth (entire) or coarsely toothed (serrate), with small bristles on the end of each tooth. Solitary or paired biennial acorns in cups about one-third the length of the nut are sessile or attached to a short pedicel. This large oak grows from Michoacán and Guerrero south to Chiapas and Guatemala, where it becomes one of the most abundant of the oaks.

Quercus emoryi Torr.
Encino Prieto

One of the most abundant oaks of northwestern Mexico, *Quercus emoryi* has lanceolate, elliptic, or obovate leaves with entire or variably toothed margins and truncated or, more commonly, cordate bases. A glossy yellow-green at maturity, the leaves when young are prominently red. Although *Q. emoryi* usually is considered deciduous, it is only after new leaves are well formed that those of the previous year fall. The leaves are glabrous on the upper surface, but somewhat puberulent along the midrib underneath. Characteristically, at the base of the leaf on the lower surface there is a prominent tuft of hairs along the midrib. The small acorns, enclosed over about one-third of their length by the small cup, mature in one year.

Q. emoryi, common along Highway 15 south of Nogales, Sonora, grows both in Sonora and Chihuahua. The edible acorns frequently are

FAGACEAE—OAK FAMILY

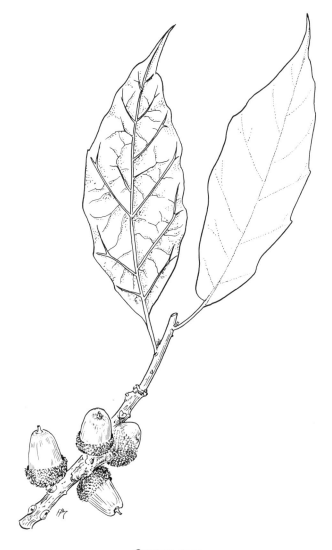

Quercus conspersa

TREE 15–20 m (50–65 ft) tall. LEAVES 12–16 cm (5–6 in) long, 3–6 cm wide, lanceolate to ovate. FRUIT 16–20 mm long, 13–20 mm wide; cup ⅓ length of acorn, sessile or on pedicel 3–10 mm long.

FAGACEAE—OAK FAMILY

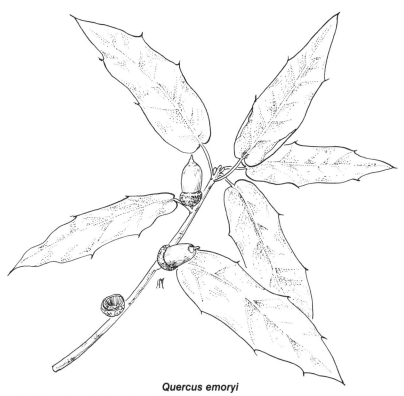

Quercus emoryi

TREE to 15 m (50 ft) tall. LEAVES 3–6 cm (1.25–2.5 in) long, 1–2 cm (.5–.75 in) wide, shape variable. FRUIT 10–15 mm long, 8 mm wide; cup 10 mm in diameter, ⅓ length of acorn, sessile.

sold in the markets under the name *bellotas*. This species was named for William Emory, a member of the commission that established the boundary between the United States and Mexico.

Quercus magnoliifolia Née
Encino Amarillo

Quercus magnoliifolia is a small tree with large, deciduous, variable leaves. Most commonly oblanceolate to obovate in shape, the leaves are rounded or truncate at the base and rounded or acute at the apex. The margins are entire, undulate, or shallowly lobed, with scurfy tomentose surfaces below and glabrous surfaces above. The similarity of the leaves to those of the magnolia prompted the specific name. Large paired or clustered acorns on prominent peduncles are enclosed over one-third of their length by the large cup. This oak is found in the western mountains from southern Sonora to Guerrero, and in the eastern mountains from Coahuila to Hidalgo and the Distrito Federal.

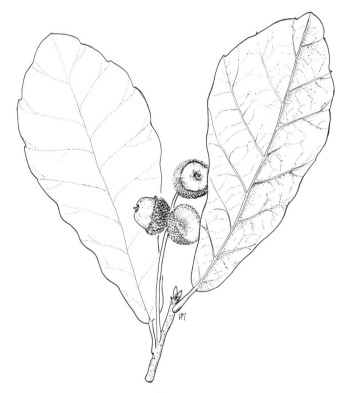

Quercus magnoliifolia
TREE 8–10 m (26–32 ft) tall, deciduous. LEAVES 10–40 cm (4–16 in) long, 5–20 cm (2–8 in) wide, oblanceolate to ovate. FRUIT 20–35 mm long, 12–22 mm in diameter; cup 15–30 mm in diameter, ⅓ length of acorn; peduncles 25–100 mm long.

Quercus pennivenia Trel.
Toché

Quercus pennivenia has large elliptic to nearly circular leaves with cordate bases and very obtuse apices. The margins are entire or minutely spined from the tips of the prominent lateral veins. Glossy and glabrous above, the foliage is dingy and stellate tomentose on the lower surface; when these hairs eventually are lost, a prominent network of veins is displayed. Axillary clusters of small acorns almost enclosed in their cups grow on short peduncles. This oak is a tree of the western Sierras from Chihuahua and Sonora south to Nayarit and Jalisco.

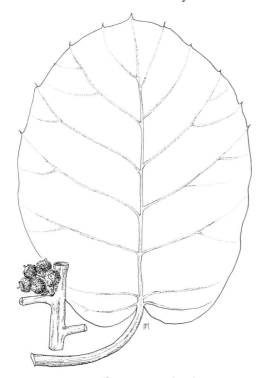

Quercus pennivenia

TREE 5–10 m (16–32 ft) tall. LEAVES 10–17 cm (4–6.5 in) long, nearly as wide, elliptic to circular. FRUIT to 15 mm in diameter, axillary clusters.

Fouquieriaceae—Ocotillo Family

Fouquieria columnaris Curran
Cirio or Boojum

A weird "upside down turnip" with small, thin branches of ocotillolike structure can be seen only in a restricted area of Baja California Norte and near Puerto Libertad, Sonora. This is the unique "cirio," or "boojum tree," which seems to be from another world. The main axis is a tall, tapering, columnar trunk with thick, whitish bark; it usually remains unbranched, except for short, lateral, twiggy branchlets and a tuft of slightly longer ones at the apex. Lateral branches are leafy and possess abundant spines; the petiole becomes a spine after the leaf falls. Subsequently, clusters of obovate or oblanceolate leaves form in the axils of the spines. The wood is soft and spongy; in larger trees it lacks the strength to keep the column erect, resulting in twisted, fantastic shapes.

At the apex of the trunk may be several panicles of small cream or yellow flowers with five overlapping sepals, five united petals, and ten exserted stamens. The fruit is a small, three-valved capsule with many seeds. Some books may cite this plant as *Idria columnaris*.

Fouquieria macdougalii Nash
Cunari or Torote Verde

Fouquieria macdougalii is a shrub or small tree with thin smooth bark peeling into paperlike bronze sheets. A loose, open top of small spiny branches, which ultimately resemble the stems of "ocotillo," develops from one to four greenish-bronze trunks. The elliptical, lanceolate or obovate leaves are structurally of two types: those on the new, long shoots are single and develop a heavy petiole, which forms a spine when the leaf falls. Those on the short shoots are in fascicles of two

---

Fouquieria columnaris
TREE to 20 m (65 ft) tall, whitish bark; spiny, twiglike branches. LEAVES 1–2 cm ▶ (.5–.75 in) long, obovate or oblanceolate, alternate. FLOWER 6–8 mm long, in terminal panicles, cream. FRUIT 12–16 mm long, yellowish capsule.

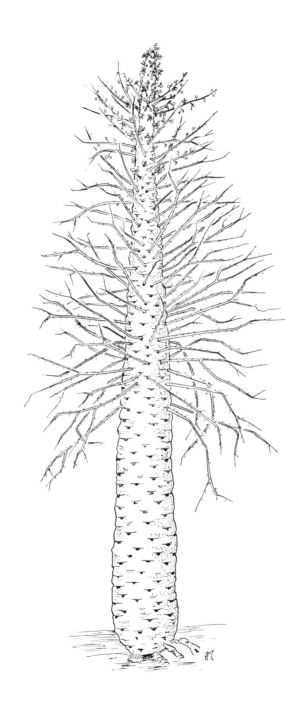

Fouquieria macdougalii

SHRUB or small tree 2–6 m (6.5–20 ft) tall; spines 4–35 mm long. LEAVES long shoot 30–55 mm long, petiole 10–30 mm long; short shoot 12–30 mm long. FLOWER 18–26 mm long, tubular, scarlet, in panicles; calyx lobes 4.5–6.5 mm long. FRUIT capsule 16–29 mm long, 5–7 mm wide.

to five leaves in the axils of the previously formed spine. The petioles of the latter leaves do not produce spines. Tubular flowers are in terminal panicles or raceme inflorescences with pink or red axes. Narrow, cylindrical, bright red corollas terminating in five erect lobes are subtended by five pink, red, or green, overlapping, ovate sepals with broad, white, papery margins. Ten scarlet stamens are strongly exserted beyond the corolla. A tan fruiting capsule contains six to thirteen winged seeds.

This species of *Fouquieria* is a member of the Sonoran desert flora and extends into the tropical deciduous forests of the Sierra Madre foothills. It ranges from central Sonora south into Sinaloa and adjacent Chihuahua. The name honors Douglas T. MacDougal, first director of the Carnegie Institution of Washington Desert Laboratory in Tucson, Arizona. In 1902 he sent a living specimen of this species to the New York Botanical Garden, where it flowered and was recognized as a new species.

Fouquieria splendens Engelm.
Ocotillo

One of the most unusual plants of the desert has long, whiplike, thorn-covered branches spreading fanlike without a trunk. This is unmistakably "ocotillo." The tall, spiny, commonly unbranched, stiff stems with gray or gray-green bark resemble dead sticks—until it rains; then clusters of shiny green leaves emerge in the axils of the spines. If the stem is elongating, the apical growth may have a single leaf on a stout petiole instead of clusters. When dryness returns the leaves are quickly shed, leaving a spine where the leaf petiole had been.

In the spring after the rains, panicles of flamelike vermillion tubular flowers develop at the ends of the branches. The tubular corolla is five-parted, with strongly recurving lobes that expose the exserted reddish-orange stamens with anthers full of bright yellow pollen. Hummingbirds find these flowers especially inviting. Upon pollination the flowers form brownish capsules, which contain seeds copiously fringed with long whitish hairs.

"Ocotillo" is a conspicuous plant of both the Sonoran and Chihuahuan deserts and of the mesquite grassland. It also is abundant on the

FOUQUIERIACEAE — OCOTILLO FAMILY

Fouquieria splendens

SHRUB 2–6 m (6.5–23 ft) tall; many branches, no trunk. LEAVES 2.5 cm (1 in) long, spatulate to obovate. FLOWER 2–2.5 cm (.75–1 in) long in panicles 5–20 cm (2–8 in) long, red. FRUIT 1–15 mm long.

shallow soil of rocky or coarse outwash slopes in Baja California, Sonora, Chihuahua, and Coahuila, south through Durango, Zacatecas, and San Luis Potosí to Hidalgo. Hedges or fences of "ocotillo" often can be seen, for cuttings stuck in the ground easily root to make a thorny barrier. This plant was named in honor of P. E. Fouquier, a Parisian professor of medicine.

Gesneriaceae—Gesneria Family

Achimenes grandiflora (Schiede) A. DC.
Patito

Of the many species of *Achimenes* growing in Mexico, *A. grandiflora* is made prominent by its showy reddish-purple flowers, borne singly in the axils of the leaves. The long slender corolla tube, which arises at an angle from the short hairy calyx, expands abruptly into five broadly spreading lobes, the lowest being the largest. Characteristic of this species is a small spur at the base of the corolla tube. Short hairs appear on the corolla tube and on both surfaces of the conspicuously toothed leaves, the veins of which are often tinged with purple or red.

Blooming from late summer into fall, this relative of the *Gloxinia* ranges from Chihuahua, Sinaloa, San Luis Potosí, Tamaulipas, and Guerrero, to Veracruz.

Kohleria deppeana Schl. & Cham.
Tochomitl

This attractive herb or shrubby plant bears its bright red or red-orange flowers in long pedunculate, axillary umbels. Three or four colorful tubular blooms radiate on their bracteate pedicels. All parts of the plant are densely pilose, with short red hairs; even the yellowish interior of the corolla is covered with bright red hairs. The tubular corolla, with its five erect or spreading, unequal lobes, arises from a much smaller five-lobed calyx, which surrounds the ovary and continues to grow as the capsule matures. Two pairs of stamens of unequal length

GESNERIACEAE — GESNERIA FAMILY

are slightly exserted beyond the corolla lobes. The paired leaves, often unequal in size, are oblong-lanceolate or oblong-ovate with crenate or dentate margins.

Growing as far north as Hidalgo and southern Nuevo Leon, *Kohleria* is primarily a plant of southern Mexico; it flourishes in Veracruz, Morelos, Guerrero, Oaxaca, and Chiapas.

Achimenes grandiflora

HERB to 60 cm (24 in) tall, perennial. LEAVES 3–12 cm (1.25–5 in) long, opposite, ovate to broadly lanceolate, acuminate, hairy. FLOWER 3.5–6 cm (1.5–2.5 in) long, tubular-funnelform, spurred; red-violet to purple. FRUIT capsule; seeds many, small.

GESNERIACEAE—GESNERIA FAMILY

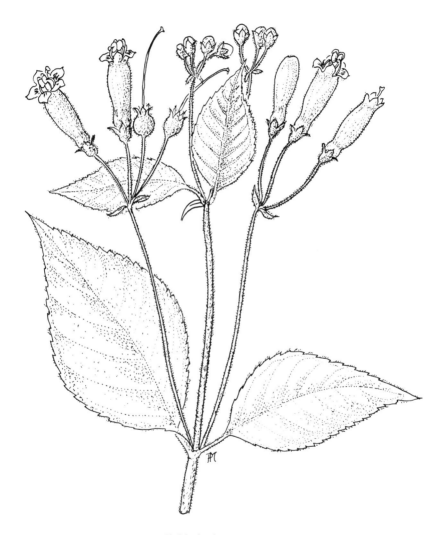

Kohleria deppeana
HERB or shrub to 2 m (6.5 ft) tall. LEAVES 5–17 cm (2–7 in) long, opposite. FLOWER 2–2.5 cm (.75–1 in) long, tubular, red; calyx 4–7 mm long; inflorescence umbellate. FRUIT capsule 6–9 mm long.

Hydrophyllaceae—Waterleaf Family

Wigandia urens Ruiz & Pav.
Ortega

Caution must be exercised in handling *Wigandia urens*, for its long hairs impart the same painful sting that comes from contact with a nettle ("ortega" means nettle). Large, showy, pale violet to purple flowers densely arranged on branched spiral (scorpioid) inflorescences attract attention to this plant. A small five-lobed calyx subtends the broad, five-lobed, shallow-funnelform corolla. The filaments of the five stamens are covered with bristles; the pistil has two prominent styles, each tipped with a broad, paddlelike stigma that remains on the hairy, mature, ovoid capsule. Large velvety leaves, cordate or truncate at the base, are irregularly serrate on the margin.

Flowering from February to October, this showy species grows from Sinaloa and Durango south to Chiapas, including San Luis Potosí and Veracruz.

Iridaceae—Iris Family

Tigridia pavonia (L.f.) Ker
Flor del Tigre or Oceloxóchitl

One of the most strikingly beautiful flowers of Mexico is the native *Tigridia pavonia*, or "tiger lily," which is not a lily but a member of the Iris Family. Arising from the corm, the cylindrical flower stalk is topped with one or two large, showy flowers. The spectacular triangular bloom is composed of three large, reflexed outer perianth segments, which rest individually on the top of the ovary. Each of these obovate

Wigandia urens
SHRUB or tree to 5 m (16 ft) tall; stinging hairs. LEAVES 5–60 cm (2–24 in) ▶ long, 3–42 cm (1.25–16 in) wide; apex rounded or obtuse; tomentose. FLOWER 1.5 cm (.75 in) long, pale violet to purple. FRUIT capsule; seeds numerous.

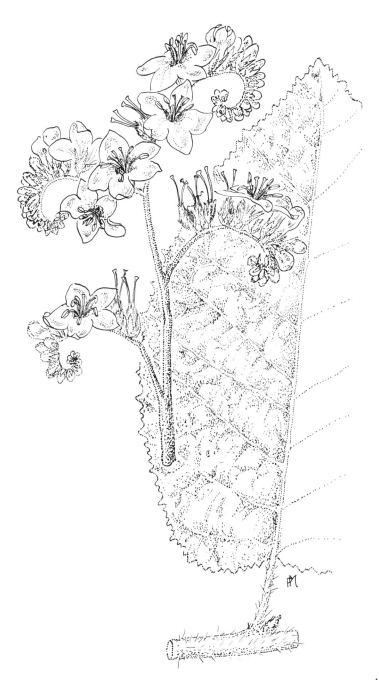

IRIDACEAE — IRIS FAMILY

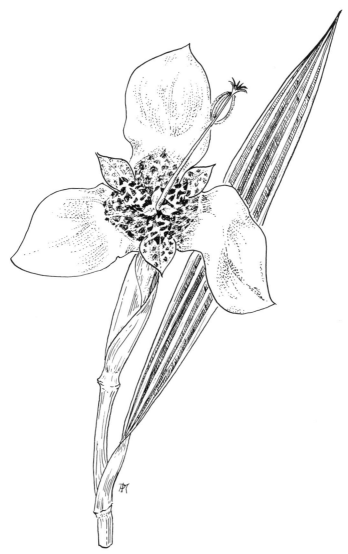

Tigridia pavonia

STEM a corm. LEAVES 40–50 cm (16–20 in) long, 1–2.5 cm (.5–1 in) wide, pleated. FLOWER 9–10 cm (3.5–4 in) in diameter; stalk 30–70 cm (12–28 in) tall; perianth orange-red, spotted. FRUIT capsule 3–5 cm (1.25–2 in) long.

segments is vivid orange-red toward the tip, becoming heavily spotted with dull red or purple toward the concave base. Resting in this cup with their tips projecting over the lip are three inner segments, about half the size of the outer perianth parts, of pale yellow; these also are copiously spotted. Standing erect from the center of the flower is a column formed by the fusion of the three stamens around the style. At the top of the column the three erect anthers arch to make a frame through which the style protrudes, with each of the three stigmas dividing into two branches. Usually only one of the fragile flowers opens at a time, each lasting but a single day. Both the long-sheathing basal leaves and the narrow lanceolate ones arising from the nodes along the flowering stalk are pleated, giving them the strength to stand erect. Each flower cluster is surrounded by one or two leaflike bracts.

Tigridia pavonia, which blooms only during the wet months, is common in the Valley of México, on the Pedregal of Mexico City, and in Jalisco, Hidalgo, Veracruz, Guerrero, and Morelos. Once one of the sacred flowers of the Aztecs and cultivated for its beauty since pre-Columbian times, it still is prized in gardens. Both the common name "flor del tigre" (flower of the tiger) and the ancient Nahuatl "oceloxóchitl" (ocelot flower) allude to the spots of the cuplike perianth.

Labiatae—Mint Family

SALVIA SPECIES

A large and complex genus of plants, *Salvia* is represented in Mexico by about 280 species of herbs and shrubs. All species have five sepals fused into a two-lipped calyx tube. In some representatives the five sepal tips can be counted; in others the fusion extends to the apex. The corollas also are tubular, ending in two lips that exhibit considerable variation in size and lobing. Fruits of all species are four one-seeded nutlets. Typical of the family Labiatae are angular stems and the presence of essential oils, which provide a wide range of flavorings from mints to sages.

Stamens of most plants consist of an anther with two adjacent pollen sacs on the end of a filament; the two stamens of *Salvia*, however, have very short filaments attached to the corolla tube. The anthers have become greatly expanded and curved, so that pollen sacs are far apart; one is near the mouth of the corolla and the other, often reduced and sterile, is near the base. With the filament as fulcrum, the anther acts as a teeter-totter; when the insect reaches for nectar at the base of the flower, it pushes the lower pollen sac up, thus causing the other sac to come down and tap the back of the insect with a dusting of pollen. Later, the stigma will extend into a position suitable for receiving pollen from the next visiting insect.

Salvia coccinea Juss.
Mirto

Characteristic of this showy red *Salvia* are long, spreading, straight hairs on the stem, calyx, and bracts, with shorter hairs on the ovate leaves. Six to ten tubular flowers, each with a prominent lower lip, are clustered at each node in the axils of lanceolate or ovate bracts to form a racemose inflorescence. Two stamens are prominently exserted beyond the short upper lip and may exceed the enlarged, lobed lower lip.

Salvia coccinea, a widely distributed species in a variety of habitats, is thought to have been introduced from Brazil. In Mexico it is most commonly found among pines and oaks in the eastern coastal states from Tamaulipas south to Yucatán, but it is also reported from Nuevo Leon, San Luis Potosí, Hidalgo, Michoacán, Oaxaca, and Chiapas.

Salvia elegans Vahl
Salvia Roja

The narrow tubular corolla of this scarlet *Salvia* opens into two long, subequal lips. The style, with unequal stigmatic lobes, projects beyond the upper lip, as occasionally do the two stamens. Flowers, forming clusters of two to six per node, arise in the axil of a deciduous ovate bract. The opposite, ovate leaves with long attenuated tips often have thin, hairy petioles that approximate the length of the blades.

A western species, *S. elegans* grows in the pine-oak forests from southern Sonora and Chihuahua along the mountain chains to Guerrero and Oaxaca. It also appears in the Trans-volcanic Cordillera in Queretaro, Hidalgo, Veracruz, and Puebla.

LABIATAE—MINT FAMILY

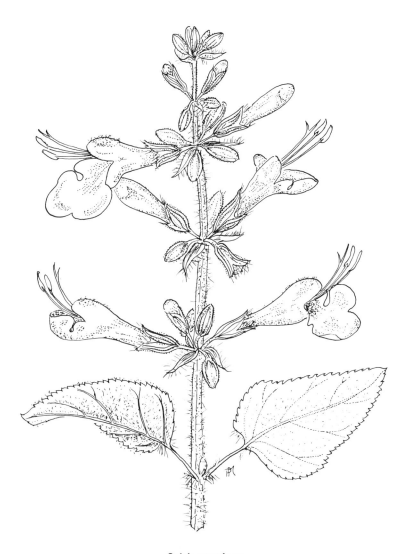

Salvia coccinea

HERB perennial or subshrub 60–100 cm (24–36 in) tall. LEAVES 2.5–7 cm (1–3 in) long, ovate, opposite. INFLORESCENCE spicate raceme; flowers 6–10 per node; corolla 25 mm long, scarlet; stamens exserted.

LABIATAE—MINT FAMILY

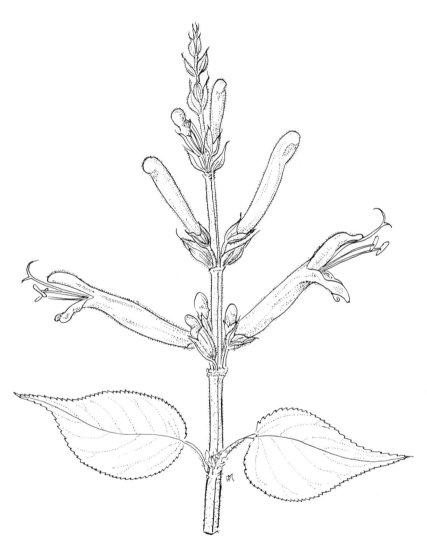

Salvia elegans

HERB perennial or subshrub, 1–2 m (3–6.5 ft) tall. LEAVES 2.5–10 cm (1–4 in) long, ovate. INFLORESCENCE 5–15 cm (2–6 in) long, spicate; flowers 2–6 per node; corolla 30–40 mm long, scarlet.

Salvia gesneraeflora Lindl. & Paxton
Salvia Roja

A large, red, inflated, globose corolla tube is typical of *Salvia gesneraeflora*. The lips of these handsome flowers are widely spreading, exposing the stamens and pubescent style under the hooded upper lip. The upper of the two extended, unequal stigmatic lobes is two or three times as long as the lower lobe and often curls over the tip of the upper corolla lip. Two to six flowers at a node are clustered in the axils of the two opposite, ovate to lanceolate bracts. Cordate-based leaves have crenate to serrate margins.

This showy species inhabits pine-oak forests in west-central and central Mexico from Durango, Jalisco, and Michoacán south through Morelos and Puebla.

Salvia greggii Gray
Autumn Sage

A short, bulbous corolla tube tipped with a broadly expanded lower lip makes this red to red-purplish *Salvia* outstanding in color and in form. A glandular-haired upper lip, smaller than the lower, is hood-shaped and hides the two stamens. One long lobe of the stigma protrudes beyond the upper lip and curls back, as if attempting to scratch the spot where an aphid might have been. Flowers, two to a node, are each subtended by a large, deeply split, two-lobed calyx. Glandular hairs cover the surfaces of the ovate, sessile leaves.

A plant of rocky hillsides of the Chihuahuan Desert, *S. greggii* is reported from Coahuila, Nuevo Leon, Zacatecas, San Luis Potosí, and Durango. This species, in cultivation since 1885, was named to honor the early frontier trader, author, and collector Josiah Gregg (1806-1850), who traveled widely in the southwestern United States and Mexico.

Salvia iodantha Fern.

Salvia iodantha is one of several *Salvia* species with rose-purple flowers. Short internodes and 10 to 12 flowers clustered at the nodes give the appearance of a very dense spicate inflorescence. The individual

LABIATAE—MINT FAMILY

Salvia gesneraeflora
HERB perennial or subshrub 2–4 m (6.5–13 ft) tall. LEAVES 5–10 cm (2–4 in) long, ovate. INFLORESCENCE 10–20 cm (4–8 in) long, raceme; flowers 4–8 per node; corolla 5+ cm (2+ in) long, red; bracts 5–10 mm.

LABIATAE—MINT FAMILY

Salvia greggii

SHRUB 30–35 cm (12–13 in) tall. LEAVES 1–3.5 cm (.5–1.25 in) long, obovate to ovate. INFLORESCENCE 5–15 cm (2–6 in) long, raceme; flower 25–35 mm long, red to purplish-red.

LABIATAE—MINT FAMILY

Salvia iodantha

SHRUB 1–3 m (3–10 ft) tall. LEAVES 3.5–12 cm (1.25–5 in) long, ovate to lanceolate. INFLORESCENCE spicate, 15–30 cm (6–12 in) long; flower 2.5 cm (1 in) long, rose-purple, 10–12 per node.

flower has a short, inconspicuous calyx subtending a narrow corolla tube. From the upper lip of the corolla, which is several times longer than the lower lip, two stamens and the style conspicuously protrude. Although variable in shape, ovate to lanceolate leaf blades commonly are supported on long thin petioles.

A western species of pine-oak forests, this *Salvia* grows from Sinaloa, Durango, and Jalisco south through Michoacán and Guerrero. It is abundant and prominent along the Mazatlan-to-Durango Highway.

Leguminosae—Pea Family

The Pea Family, one of the three largest families of flowering plants, is well represented throughout Mexico. Leguminosae commonly is divided into three subfamilies based on differences in flower structure, even though all members have a legume-type fruit. The Mimosa subfamily (Mimosoideae) has very small flowers clustered into tight heads or spikes. As the petals, which are of equal size, are inconspicuous, the showy structures are the stamens. In the Caesalpinia subfamily (Caesalpinioideae), with flowers in various types of inflorescences, the five petals are larger and of unequal size. The upper petal is innermost in the bud, constituting an important distinction from the Papilio subfamily (Papilionoideae), in which the upper petal (banner), commonly the largest, is outermost in the bud and tends to enclose the other four. The two lateral petals are termed wings, and the two remaining form the keel that surrounds the stamens and pistil. Papilio flowers, typically shaped like those of the sweet pea, are arranged in a variety of inflorescence types.

CAESALPINIOIDEAE SUBFAMILY

Caesalpinia cacalaco Humb. & Bonpl.
Cascalote

In the winter months, a spiny shrub with spikes of lemon-yellow flowers or long reddish seed pods is noticeable in many areas of Mexico. The large, delicate flowers with five free yellow petals form long, dense,

LEGUMINOSAE—PEA FAMILY

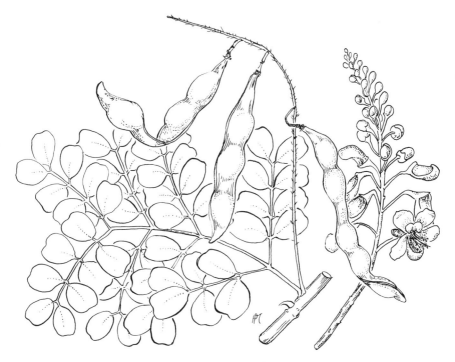

Caesalpinia cacalaco
SHRUB or tree often spiny. LEAVES bipinnate; leaflets 1–2.5 cm (.5–1 in) long. FLOWER 2.5 cm (1 in) long, in long racemes, yellow. FRUIT pod 10–15 cm (4–6 in) long.

showy racemes. Reddish, linear seed pods, which are somewhat fleshy and cylindrical with constrictions between the seeds, persist on the plant long after the flowers. When *Caesalpinia* is a large tree, the bark is rough and gray. Leaves on both the shrub and tree are bipinnate, with few oval or suborbiculate, opposite leaflets.

This species is a native of the tropical deciduous forests from Sinaloa to Oaxaca and Veracruz, but it has been spread outside its natural area by cultivation and commercialization. The fruits, which contain 25 to 30 percent tannin, have been used locally and exported for use in tanning. Early inhabitants obtained a black dye from the seed pods and a red dye from the wood.

Caesalpinia crista L.
Ojo de Venado

"Ojo de venado" (eye of the deer), so named for its large gray seeds, is a straggling shrub covered with numerous straight or curved prickles. Bipinnately compound leaves have ovate to orbiculate leaflets. Small greenish-yellow flowers borne in dense racemes produce seed pods as broad as they are long and densely covered with sharp spines. Each fruit produces two very hard, shiny, gray seeds nearly an inch in diameter.

This *Caesalpinia* is a common strand plant, where it may form dense thickets. It occurs from Sinaloa southward on the west coast and from Veracruz on the east coast. Its wide distribution along the shores of the tropical seas may have resulted from the ability of the seeds to remain viable for extended periods while being transported by ocean currents. Children use the seeds as marbles, and they serve as a source of medicine.

Caesalpinia pulcherrima (L.) Swartz
Tabachín or Mexican Bird-of-Paradise

Clusters of bright scarlet flowers with long red stamens characterize *Caesalpinia pulcherrima*, a plant not to be confused with *Delonix regia*, the "royal poinciana." In *Caesalpinia,* individual flowers are smaller, as are the seed pods. Each flower in the terminal or axillary raceme has five red sepals, the lower one enlarged and hoodlike, enclosing the other flower parts in bud. The clawed, separate petals are vermillion red, often edged or variegated with yellow; the upper petal is smaller than the others. Ten bright red, conspicuous stamens and a single pistil extend gracefully beyond the corolla lobes. Flattened, reddish-brown, persistent seed pods are explosively dehiscent, propelling seeds to a great distance. The stems, unarmed or with a few weak prickles, have alternate, bipinnately compound leaves with six to twelve pairs of pinnae, each bearing a like number of oblong to ovate leaflets. Crushed leaves exude a disagreeable odor.

"Tabachín" is distributed over much of Mexico as a cultivated shrub or small tree. In many areas it has escaped from cultivation and become established as a native. This showy shrub, in addition to its

Caesalpinia crista
SHRUB at seashore, spiny. LEAVES bipinnately compound; leaflets 1.5–4 cm (.75–1.5 in) long. FLOWER greenish-yellow, small. FRUIT spiny, 6–8 cm (2.5–3 in) long.

ornamental value, has other uses: its sweet-scented flowers yield a good quality honey, tannin is obtained from the fruits, its seeds are cooked and eaten when green, and various plant parts are used in medicines.

The "Mexican bird-of-paradise" is closely related to the "yellow bird-of-paradise," *C. gilliesii* Wall., a plant of northern Mexico recognized by its copious pubescence and glandular inflorescences, characters absent in *C. pulcherrima*. "Bird-of-paradise" also is applied to other, unrelated plants.

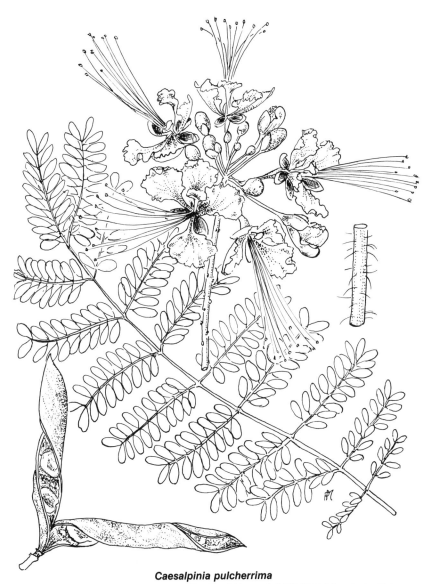

Caesalpinia pulcherrima

SHRUB or small tree 1−6 m (3−20 ft) tall; prickles soft. LEAVES bipinnate; leaflets 1.5−2.5 cm (.5−1 in) long. FLOWER 4−7 cm (1.5−3 in) wide, red or red and yellow; stamens 5−7 cm (2−3 in) long, red. FRUIT 10 cm (4 in) long, 2 cm (.75 in) wide.

LEGUMINOSAE – PEA FAMILY

Cassia biflora L.
Flor de San José or Abejón.

Blooming most of the year, *Cassia biflora* colors the landscape with its bright yellow flowers, which grow profusely in axillary pairs (*biflora* = two flowers). The five petals of each flower are of unequal shape and size, as are the ten stamens, which with an elongated curved pistil extend prominently from the center of the flower. Green, immature seed pods of all sizes are found on the shrub as soon as the flowers are pollinated, but the mature fruit is a narrow, linear, dark brown pod with corded margins. The compound leaves have from four to eight pairs of leaflets, with a conspicuous gland between the lowest pair.

This species is widely distributed from Baja California to Chihuahua, and south to Yucatán and Chiapas. Specimens may be found in flower at any time of the year, but the peak season is October to April.

Cassia tomentosa L.f.
Retama

Bright yellow flowers in terminal or axillary clusters will attract attention to this shrub or small tree. The five obovate petals creating a large cup-shaped flower are of almost the same shape and form. From the center of the blossom project the long, hairy, curved pistil and 10 stamens with anthers of three different sizes. As is typical of members of this genus, the seven fertile anthers open by terminal pores. The pistil develops into a narrow, elongated, somewhat compressed fruit. The stems, leaves, leaflets, and fruits are densely covered with hairs (*tomentosa* = hairy), although the leaflets may be more tomentose on the lower surface than on the upper. The pinnately compound leaves are composed of 12 to 16 leaflets, each with a small point.

Cassia is a genus of many species; "retama," like several others, frequently is cultivated for its ornamental value. It occurs naturally from Queretaro and Hidalgo south into Guatemala.

LEGUMINOSAE — PEA FAMILY

Cassia biflora

SHRUB to 5 m (16 ft) tall. LEAVES pinnate; leaflets 4–10 pairs, 1–3.5 cm (.5–1.5 in) long, ovate to elliptic. FLOWER 5 cm (2 in) in diameter, paired, yellow. FRUIT 5–15 cm (2–6 in) long, 4–5 mm broad, linear, thin.

LEGUMINOSAE — PEA FAMILY

Cassia tomentosa

SHRUB or tree 3–7 m (10–23 ft) tall. LEAVES pinnate; leaflets 2–6.5 cm (.75–2.5 in) long, 6–8 pairs, oblong. FLOWER 2.5–4 cm (1–1.5 in) wide, terminal or axillary clusters, yellow. FRUIT 8–12 cm (3–5 in) long.

LEGUMINOSAE — PEA FAMILY

Cercidium praecox (Ruiz & Pav.) Harms
Palo Mantescoso or Palo Verde

Yellow is one of the most common flower colors of desert plants, but there is nothing ordinary in the sight of blooming "palo verde" trees in the spring. Each brilliant flowering tree, with its many ascending or spreading branches completely covered with golden blossoms, is a new enjoyment. *Cercidium praecox* is one of five species of *Cercidium* in Mexico. The flowers have five almost equal, spreading petals with 10 erect stamens to add more color. A slightly larger upper petal is dotted with small red spots, which aid in distinguishing it from other species of *Cercidium*. The bark of both the branches and trunk is lime-green, smooth, and unmarred. Bipinnate leaves may develop a short spine. Prior to the development of new leaves, the raceme inflorescences form in clusters of one to three at a node and bear one to six distinctly pedicelled flowers. Thin, papery, tan legumes with one or two seeds are not constricted.

This species of *Cercidium* grows in Sonora, Sinaloa, and southern Baja California. Four other species of "palo verde" in Mexico produce similar brilliant floral displays: *C. texanum* is found in the northeastern states of Coahuila, Nuevo Leon, and Tamaulipas; *C. floridum*, a species of Sonora, Sinaloa, and Baja California, has inflorescences on terminal and subterminal branches and flattened legumes not strongly tapered at the apex. *C. microphyllum* lacks the spines of the other species, but each branchlet becomes sharply pointed and spinelike, with small leaves (*micro* = small; *phyllum* = leaf) and yellow-green bark. Its cylindrical pods are prominently constricted between the seeds and strongly tapered at each end. This species grows from Arizona and southeastern California south through Sonora and Baja California.

Delonix regia (Boj.) Raf.
Tabuchín, Flamboyán, or Royal Poinciana

A crown of bright scarlet flowers on a leafless tree or against a background of feathery green leaves signals one of Mexico's most magnificent trees, the "royal poinciana." The showy blooms are in loose racemose inflorescences. Each of the five brilliant petals tapers into a

LEGUMINOSAE—PEA FAMILY

Cercidium praecox

SHRUB or tree 2—9 m (6.5—30 ft) tall; bark lime-green; spines 2—18 mm long. LEAVES bipinnate; pinnae 1 pair; leaflets 3—10 mm long, 5—9 pairs per pinna. FLOWER yellow, 1—6 per raceme; pedicels 7—10 mm long. FRUIT 3—6 cm (1.25—2.5 in) long, 6—10 mm wide.

narrow base (claw). Filling the space between the claws are five decorative calyx lobes, which are red inside and green outside with a yellow border. Ten projecting, prominent, red stamens tipped with orange anthers curve upward from the open flower.

Following these showy flowers are conspicuous long, dark brown, flattened seed pods, which produce numerous hard seeds. At maturity the pod is hard and woody, becoming a rattle when the enclosed seeds loosen from the placenta. The long dark fruits contrast with the deciduous light green, fernlike, bipinnate leaves with many small leaflets.

LEGUMINOSAE — PEA FAMILY

This distinctive tree is not a native of Mexico, but comes from Madagascar and tropical Africa. It was named for Poinci, Governor of Antilles and a patron of botany in the mid-seventeenth century. It is widely used as an ornamental street tree, but in many places it has become established without cultivation.

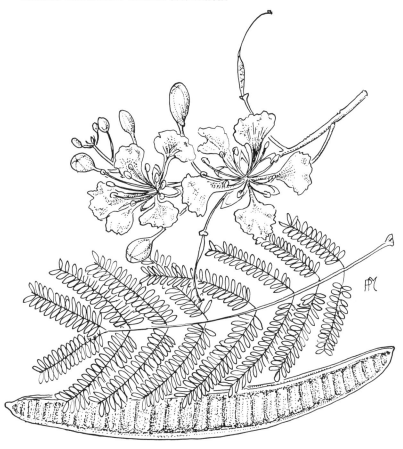

Delonix regia
TREE to 20 m (65 ft) tall; crown spreading. LEAVES bipinnate; pinnae 10–20; leaflets 20–40, small, 4–10 mm long. FLOWER scarlet; petals 6–7 cm (2.5–3 in) long. FRUIT to 60 cm (24 in) long, flattened.

LEGUMINOSAE — PEA FAMILY

Haematoxylon brasiletto Karst.
Brasil

This showy shrub or small tree has profuse racemes of small yellow flowers, which are produced in the axil of a leaf or from nodes where previous leaves were borne. Alternate, evenly pinnate compound leaves usually have four to eight leaflets, but occasionally on new growth the leaves are twice compound. Young leaflets are bronze-tinged, obovate, with a notch at the apex. The five oblong-ovate petals of the corolla are unequal in size and widely spreading. All are pale yellow and tinged with red at the base; one has additional red streaks. Ten distinct stamens stand erect from the slightly recurved corolla. The five-lobed calyx tube is conspicuous as a shallow bowl beneath the maturing fruit. The flat pods are a deep red-brown and have a faint lateral line, which will open for seed dispersal.

"Brasil" can be found in relatively dry locations throughout western Mexico, from Sonora and Baja California to Oaxaca, and is easily recognized by its reddish-brown bark on young trees, or deeply fissured gray bark on older stems. The heartwood turns dark red on exposure to the air and is a source of a red dye used for coloring cotton cloth.

Hymenaea coubaril L.
Cuapinol

Either the peculiar compound leaves, consisting of two leaflets with noticeably unequal sides, or the large, dark brown pods make *Hymenaea* a conspicuous tree. It often grows so tall that it may produce buttresses for additional support. The shiny, green, paired, asymmetrical leaflets are elliptic ovate in shape and slightly leathery in texture. From March to July flowers arranged in erect, terminal, panicle-type clusters have five white, elliptic petals with brown tips, 10 stamens, and a stalked pistil. Subtending the corolla is a bell-shaped, finely hairy, thick basal cup with four or five gray-green, hairy sepals. The erect fruit, a rough pod with thick walls, does not open and often remains on the tree after maturity. It contains a mealy, yellow pulp surrounding red, green, or

LEGUMINOSAE — PEA FAMILY

Haematoxylon brasiletto
SHRUB or tree 2–5 m (6.5–16 ft) tall, spiny or spineless. LEAVES 0.5–2.5 cm (.25–1 in) long, pinnate. FLOWER 0.5 cm (.25 in) long, yellow, axillary racemes. FRUIT 3.5 cm (1.5 in) long, red-brown.

LEGUMINOSAE—PEA FAMILY

Hymenaea coubaril
TREE 5–30 m (16–100 ft) tall, evergreen. LEAVES compound, 2 asymmetrical leaflets 5–10 cm (2–4 in) long. FLOWER 3.5 cm (1.25 in) in diameter, white. FRUIT thick, 10–17.5 cm (4–7 in) long, 4–6.5 cm (1.5–2.5 in) broad.

LEGUMINOSAE — PEA FAMILY

dark brown seeds. The sweet pulp is edible but has an unpleasant odor. The fruits occasionally are sold in the markets.

"Cuapinol," sometimes spelled "guapinol," is found along the east coast from southern Veracruz to Campeche. Along the Pacific coast it extends from Nayarit to Chiapas and south into Central America.

MIMOSOIDEAE SUBFAMILY

Acacia farnesiana (L.) Willd.
Huisache or Vinorana

When in bloom, the sweet fragrance of countless tiny flowers discloses the presence of *Acacia farnesiana* before it comes into view. Almost immediately after the delicate new leaves emerge in the spring, the spreading flexible branches are completely covered with bright yellow balls made up of numerous small flowers, each with many exserted stamens. The fluffy balls, feathery leaves, and graceful branches accented with dark brown bark make a beautiful tree. The bipinnate leaves consist of two to six pairs of pinnae, each with many tiny leaflets. The dark brown or purplish fruit, about as long as the leaves, is a hard, woody, cylindrical legume tapering at both ends.

A native of the New World, but now very common and widely distributed throughout the tropics, *A. farnesiana* has been cultivated as a source of fragrant oils for perfumes.

Acacia greggii Gray
Uña de Gato or Cat-claw Acacia

Fluffy, spikelike clusters of numerous, small, creamy-white flowers make this a showy shrub or small tree in the spring. It grows commonly as a shrub, but in favorable locations it may become a tree. Sharp, recurved spines give rise to the common name "uña de gato," or "cat-claw acacia." The alternate leaves are bipinnately compound with one to three pairs of pinnae, each having three to seven pairs of leaflets. The fruits are flattened, twisted pods, which turn from light green to reddish brown upon maturity.

LEGUMINOSAE—PEA FAMILY

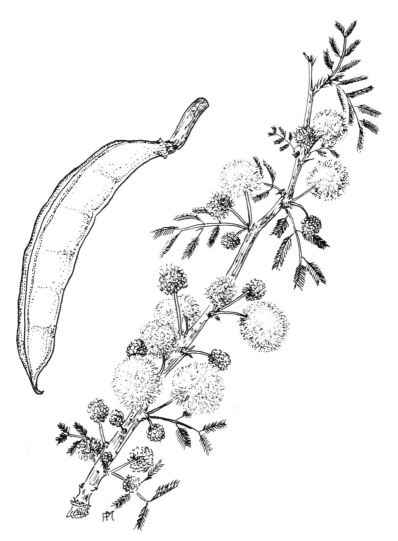

Acacia farnesiana

SHRUB or tree 1–9 m (3–30 ft) tall; spines 1–2.5 cm (.5–1 in) long. LEAVES 3–8 cm (1.25–3.25 in) long, bipinnate; pinnae 2–6; leaflets 2–5 mm long. FLOWER heads 1 cm (.5 in) in diameter, capitate, yellow. FRUIT 3–8 cm (1.25–3.25 in) long, cylindrical, dark brown or purple.

LEGUMINOSAE—PEA FAMILY

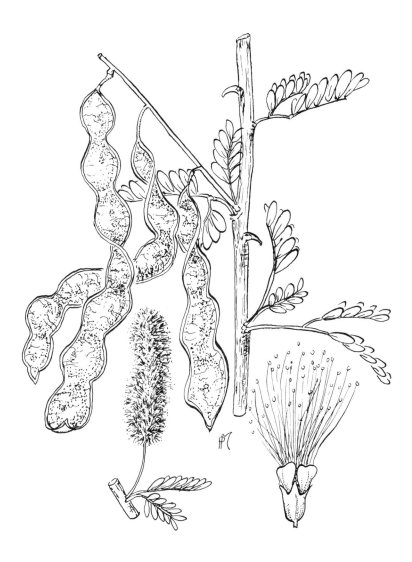

Acacia greggii

SHRUB or tree 1–8 m (3–26 ft) tall; spines 3 mm long, sharp, curved. LEAVES bipinnate; pinnae 1–1.5 cm (.5 –.75 in) long; leaflets 4–6 mm long. FLOWER creamy white, in spikes. FRUIT 5–7.5 cm (2–3 in) long, flat, twisted.

This common plant has many uses: a mush or meal can be made of the seeds, the flowers yield important amounts of honey, and cattle eat the new spring growth. *A. greggii*, found in deserts and spiny forests of northern Mexico from Tamaulipas to Baja California, also grows north of the border from Texas to California.

* * *

Large-thorned acacias. Several Mexican species of *Acacia* are noteworthy for their large, swollen spines. *A. cochliacantha* and *A. cornigera* are discussed here. Care must be taken in examining these plants because the hollow spines often are inhabited by colonies of ants. By biting and stinging, these ants protect the plant from herbivores, mainly insects. They also clean out intruding vines that would shade the acacia leaves. The plants symbiotically have developed glands along the petiole and leaf axis that secrete nectar, primarily carbohydrates, which the ants eat. In addition, the leaflet tips are modified structures which ants harvest for the proteins, fats, and carbohydrates necessary to nourish ant larvae.

Acacia cochliacantha Humb. & Bonpl. ex Willd.
Quisache Tepamo or Boat-spined Acacia

Large, gray or brown, boat-shaped spines, especially on sterile branches, are a conspicuous mark of this shrub or small tree, with gray to almost black, deeply fissured bark. Numerous small leaflets make up the feathery, bipinnate leaves. Flowers are numerous, small, yellowish, and clustered in fuzzy, rounded heads that frequently are paired at each node. The dark brown fruits are tardily dehiscent.

Acacia cochliacantha occurs from Baja California, Sonora, and Chihuahua south to Veracruz and Chiapas. It frequently is a weedy, low shrub lining the roadside. Flowering time is most commonly from March to May, but there may be a second flowering from July to November.

LEGUMINOSAE — PEA FAMILY

Acacia cochliacantha
SHRUB or tree 1.5–4.5 m (5–14 ft) tall. LEAVES bipinnate; leaflets 1–1.5 mm long. SPINES 1.5–5 cm (.75–2 in) long, boat-shaped. FLOWER heads 6–7 mm diameter, yellowish capitate. FRUIT 7.5–10 cm (3–4 in) long.

LEGUMINOSAE — PEA FAMILY

Acacia cornigera (L.) Willd.
Árbol de Cuerno, Cuerno de Toro, or Bull's Horn Acacia

Large, round, hollow spines looking much like the horns of a bull mark the small shrub or tree *Acacia cornigera* (*cornigera* means hornbearing). These peculiar, brownish-yellow or black appendages, paired at each leaf, are partially united at the base. The leaves, which are twice compound, have many small, glabrous leaflets. Along the petioles are "canoe-shaped" nectar glands. Abundant small yellow flowers with numerous exserted stamens bloom in dense, thick, cylindrical spikes. Eventually produced are red or brown fruits, abruptly contracted into a prominent, needlelike beak at the apex. The seeds, in an indehiscent pod, are surrounded by a sweet edible pulp, which invites birds and humans to scatter the seeds as the fruit is eaten.

Acacia cornigera grows from southern Tamaulipas south through Veracruz to Oaxaca and Chiapas. It is common on the Yucatán Peninsula and along the Pacific slope of Guerrero.

Calliandra eriophylla Benth.
Huajillo, Charresquillo, or Fairy Duster

There are relatively few flowers per head on *Calliandra*, but the shrub is made conspicuous by the numerous, showy, long, pink stamens exserted beyond the small, reddish-purple, inconspicuous petals. These delicate blooms give rise to the popular name "fairy duster" in the southwestern United States. The shrub is low and bushy, with gray bark and small bipinnate leaves. Each member of the two or three pairs of leaflets (pinnae) will bear seven to nine pairs of minute, secondary leaflets. Flat, velvety, gray seed pods are conspicuous for their thickened, cordlike margins.

Calliandra can be found in the desert and grassland areas from Baja California to Coahuila and south as far as Puebla.

Acacia cornigera

SHRUB or tree to 6 m (20 ft) tall. SPINES 2.5–10 cm (1–4 in) long, paired, ▶ "bull-horn"-shaped. FLOWER yellow, in spikes. FRUIT 2.5–6 cm (1–2.5 in) long, red or brown.

LEGUMINOSAE — PEA FAMILY

LEGUMINOSAE — PEA FAMILY

Calliandra eriophylla

SHRUB low, bushy, to 90 cm (35 in) tall. LEAVES bipinnate; leaflets 3–4 mm long. FLOWERS in small heads; stamens prominent. FRUIT 7.5 cm (3 in) long, flat.

LEGUMINOSAE—PEA FAMILY

Calliandra houstoniana (Mill.) Standl.
Barbas de Chivato or Beard of the Goat

This species may become a small tree, although most commonly it is a shrub. Feathery, bipinnate leaves bear seven to twelve pairs of primary leaflets (pinnae), each with 30 to 40 pairs of secondary leaflets. Flowers borne in small heads on terminal racemes have brownish, hairy corollas. The numerous, spectacular, red-purple stamens make this shrub conspicuous when in bloom. Thick, woody, reddish-brown seed pods with heavy marginal cords also add color to the plant.

Sonora to Tamaulipas and south to Veracruz, Chiapas, and on to Guatemala is the range of this species. Similar to *C. houstoniana* is *C. anomala* ("cabellos de angel") which differs in having 15 to 20 pairs of pinnae with more numerous leaflets per pinna, and other more subtle differences; both species have the same distribution.

Enterolobium cyclocarpum (Jacq.) Griseb.
Orejón or Guanacaste

Enterolobium immediately attracts attention because of its large size and shape and its distinctive seed pods. The broadly spreading crown usually exceeds the height, and it frequently is planted for shade. Glabrous, fernlike, green leaves are bipinnate with four to nine pairs of pinnae, each with numerous pairs of leaflets. In winter, many inconspicuous white flowers are clustered into ball-like heads. Unusual flat seed pods curled into a circle, persist on the tree or cover the ground (Greek *cyclos* = wheel or circle, *carpos* = fruit). Young fruits are shiny green but mature into a polished dark brown or black. The thinness of the pod makes the position of the several seeds conspicuous. Toasted seeds are edible, as are the green pods; cattle eat the mature pods. Both the pods and the rough bark are rich in tannin.

The soft, light wood is used in carpentry, furniture, and veneers. Indians found the large trunks suitable for dugout canoes. *Enterolobium* is reported from southern Baja California and Sinaloa to Tamaulipas and south into Central America.

LEGUMINOSAE — PEA FAMILY

Calliandra houstoniana

SHRUB or small tree 1–5 m (3–16 ft) tall. LEAVES bipinnate; leaflets 4–7 mm long. FLOWERS in terminal spike or raceme; petals 10 mm long; stamens 5 cm (2 in) long, prominent. FRUIT 12 cm (5 in) long.

LEGUMINOSAE – PEA FAMILY

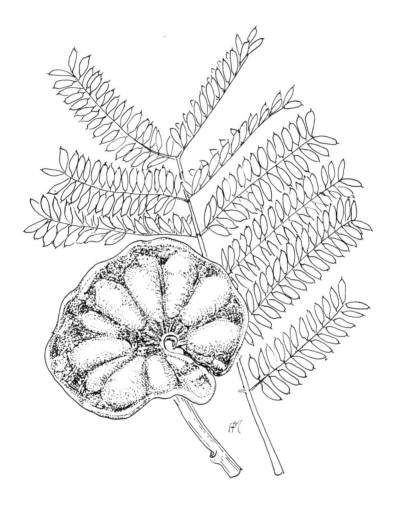

Enterolobium cyclocarpum

TREE 18−30 m (60−100 ft) tall, unarmed, crown 23−45 m (75−135 ft) broad; trunk 1−2.5 m (3−7 ft) in diameter, rough. LEAVES bipinnate; pinnae 4−9 pairs; leaflets 10−12 mm long, numerous. FLOWER capitate, sessile, white. FRUIT 8−11 cm (3−4 in) in diameter, coiled, flat.

LEGUMINOSAE—PEA FAMILY

Lysiloma divaricata (Jacq.) Macbr.
Quiebracha

Another of the large, feathery-leaved leguminous trees, *Lysiloma* is conspicuous when flowering, with its numerous balls of small white flowers with prominent stamens, or in fruit, with its long, brown, flattened pods. Typical of the genus, the fruits have marginal ribs that separate from the body of the pod, freeing the walls and allowing the seeds to drop out. The evergreen leaves are bipinnate with four to nine pairs of pinnae, each having numerous pairs of small leaflets. On new growth, two prominent stipules with asymmetrical bases subtend each new leaf.

A tree of central Sonora and Baja California, this species grows as far south as Oaxaca and Veracruz.

Pithecellobium dulce (Roxb.) Benth.
Guamúchil

The curious, irregularly curved and coiled fruits of *Pithecolobium* decorate this shrub or small tree as they hang in numbers from the elongated racemes. Before fruiting, the long pendant inflorescences consist of numerous ball-like clusters of small white or cream-colored flowers with conspicuously exserted stamens. Bipinnate leaves with only two pinnae, each with two obovate leaflets, are persistent on the tree until new leaves begin to form. One or two short stipular spines may develop at the base of the leaves. As a broadly spreading tree, *Pithecellobium* frequently is planted for shade, as well as for the edible, fleshy covering surrounding each of the seeds. The young pods may also be eaten, and the leaves furnish forage for cattle.

Common in frost-free areas, this species flowers from November to June in Baja California, Sonora, Chihuahua, and Tamaulipas south to Chiapas and Central America. It has been introduced into the Philippines and India, where it is cultivated for its food and forage value. Tannin is obtained from the bark.

LEGUMINOSAE—PEA FAMILY

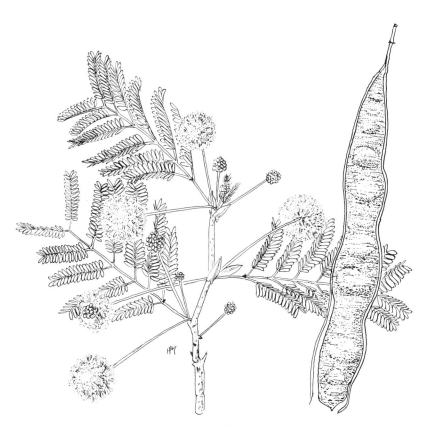

Lysiloma divaricata
SHRUB or tree to 15 m (50 ft) tall. LEAVES bipinnate; pinnae 4−9 pairs; leaflets 20−30 pairs, 3−8 mm long; stipules ovate. FLOWERS capitate, white; stamens prominent. FRUIT 8−15 cm (3−6 in) long, 1−2 cm (.5−.75 in) wide, marginal ribs detaching.

LEGUMINOSAE—PEA FAMILY

Pithecellobium dulce

SHRUB or tree to 15 m (50 ft) tall, armed. LEAVES bipinnate, pinnae 1 pair; leaflets 2–7 cm (.75–2.75 in) long, obovate, 1 pair. FLOWERS in racemes to 40 cm (15 in) long, cream to white. FRUIT 10–18 cm (4–7 in) long, curved and coiled.

LEGUMINOSAE — PEA FAMILY

Prosopis juliflora (Swartz) DC.
Mesquite or Algarroba

Omnipresent throughout much of Mexico, the "mesquite" tree brings welcome shade, beauty, and a touch of green to desert areas from spring through fall. The light green of the feathery bipinnate leaves contrasts with its dark brown or blackish trunk and frequently armed branches. During the spring, pendant yellow-greenish spikes of small fragrant flowers attract bees, which serve as pollinators. Edible, light tan to brown pods filled with beans hang from the branches. Contrary to the usual fruit development of the Pea Family, the long narrow pods do not split lengthwise, but break into segments for seed dispersal. The fruits have a high sugar content and may serve as food for beetles, cattle, and rodents. The mature pods and their enclosed beans were ground into flour by early human residents.

"Mesquites" are exceptionally useful plants; besides the edible foliage, pods, seeds, and honey-producing flowers, the hard wood is of lumber and furniture quality, burns slowly as firewood, and resists decay as fencing material.

The several species of *Prosopis* are widely distributed and difficult to distinguish one from another. According to one specialist, *P. juliflora* is a west coast species growing from Sinaloa to Oaxaca. In the north central and eastern states it is replaced by *P. glandulosa*, which differs in the length-to-width ratio of the leaflets. In Baja California and Sonora *P. velutina* is a common species, which differs from *P. juliflora* by having pubescent instead of glabrous leaves.

PAPILIONOIDEAE SUBFAMILY

Andira inermis (Swartz) H.B.K.
Cuilumbuca

Showy panicles of fragrant, pink to purple, pea-shaped flowers make *Andira* an attractive tree from February through April. The smallness of the individual flowers is compensated by the profusion of blooms on long terminal or axillary panicles. A tiny, minutely toothed calyx of dark purple contrasts with the purplish pink petals of the small flowers.

LEGUMINOSAE — PEA FAMILY

As is common in many of the legumes, there are five petals of unequal size; in *Andira* the largest petal—the broad, rounded banner—is rosy purple, while the smaller, narrower wings and keel are a lighter pinkish

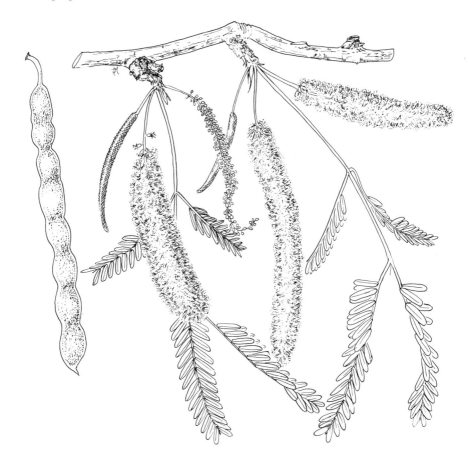

Prosopis juliflora

SHRUB or tree to 12 m (40 ft) tall, bark dark brown, armed. LEAVES bipinnate; pinnae 1—4 pairs, 3—11 cm (1.25—4.25 in) long; leaflets 11—15 pairs, 6—23 mm long. FLOWERS in spikes 7—15 cm (3—5 in) long, greenish-white or pale yellow. FRUIT 8—29 cm (3—11 in) long, 9—17 mm wide, tan.

LEGUMINOSAE—PEA FAMILY

purple. Unlike most legumes, its fruit is a globular or ovoid drupe, with a somewhat fleshy outer coat and a hard inner layer surrounding a single seed containing a poisonous alkaloid. Large, glabrous, pinnately compound, deciduous leaves have oblong leaflets, paired except at the apex. At times, the large trunk produces support buttresses up to 3 m high.

Andira may be seen from Michoacán to Chiapas, in Yucatán and Tabasco. It grows well in wet forest areas but is not restricted to them. The abundant, fragrant flowers attract bees, other insect pollinators, and hummingbirds. Fallen blossoms form a colorful carpet beneath the tree.

Andira inermis
TREE to 25 m (82 ft) tall; bark rough, light gray. LEAVES pinnate, 15−40 cm (6−15 in) long; 9−11 leaflets 5−8 cm (2−3.25) long. FLOWER 1 cm (.5 in) long, pink to purple; panicles 15−30 cm (6−12 in) long. FRUIT 3.5−4 cm (1.25−1.5 in) long, oval, fleshy on outside.

Coursetia glandulosa Gray
Cousamo or Samo Prieto

Coursetia is a widely distributed shrub or, occasionally, a small tree, with showy white and yellow or pink and yellow flowers during March and April. As in the typical pea flower, the five petals are of unequal size, with one larger upper petal (banner), two lateral wings, and two keel petals that enclose the pistil and 10 stamens. Three to six flowers are arranged in a short, glandular, pubescent raceme, with one to two racemes on short lateral branches at the nodes. The brown seed pods are constricted between the two to twelve seeds, giving the appearance of a string of beads. Pinnately compound oblong to elliptic leaflets commonly are tipped with a small apical point; a terminal leaflet usually is absent.

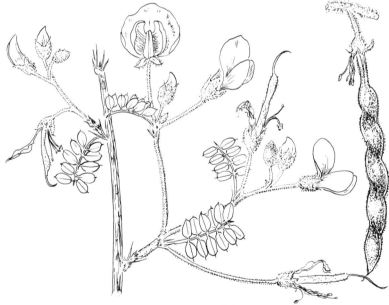

Coursetia glandulosa

SHRUB or tree 1.5−6 m (4.5−20 ft) tall, unarmed. LEAVES odd or even pinnate 2−5 cm (.75−2 in) long; leaflets 4−9 pairs, 5−15 mm long. FLOWER 11−13 mm long, yellow; racemes 2−4 cm (.75−1.5 in) long, 3−6 flowered. FRUIT 4−7 cm (1.5−2.75 in) long, 5−7 mm wide, constricted between seeds.

Coursetia favors the decomposed granite soil between boulders, on hillsides, and in arroyos in the states of Sinaloa and Sonora, as well as on the Baja Peninsula.

Erythrina flabelliformis Kearney
Colorín or Corcho

The large, terminal, cone-shaped racemes of brilliant red, pointed flowers of *Erythrina* make it conspicuous in the landscape. Flowering during the leafless period, the bloom consists of a long, rolled upper petal extending beyond the campanulate calyx to cover four lower, clawed petals. Also concealed beneath this elongated banner petal is the staminal column of nine partially fused stamens and one free stamen. At maturity, an increased accentuation of the curve of the banner reveals the tips of the stamens and anthers. The flower is followed by a long woody pod, pendant on a short stalk and ending in a long apical point. At maturity the pod splits along one side, exposing one to many bright red or orange, hard, highly poisonous seeds. Pinnately compound leaves of three broad triangular leaflets appear after the flowers. Cat-claw shaped spines are intermittently spaced along some of the petioles and stems.

This frost-sensitive native of desert grasslands and oak woodlands grows on rocky slopes and washes from Baja California, Sonora, and Chihuahua south through Durango, Zacatecas, and San Luis Potosí to Jalisco, Michoacán, and Morelos.

Gliricidia sepium (Jacq.) Steud.
Cacahuananche, Mata Ráton, or Madre Cacao

Large clusters of racemes bearing bright pink to white pealike flowers appear from December to April on the branches of the leafless *Gliricidia* tree, lending it the appearance of a fruit tree in bloom. The abundant delicate flowers, shaped like miniature sweet peas, grow on short pedicels. Emerging later are pinnately compound leaves composed of seven to seventeen ovate to elliptic leaflets, which when young are tinged with bronze. Feathery leaves and black, woody, legume fruits follow the flowers from February to July.

Erythrina flabelliformis
SHRUB or tree armed; bark gray. LEAVES pinnate, deciduous; 3 leaflets 2.5–7.5 cm (1–3 in) long, 3.5–11 cm (1.5–4.25 in) broad. FLOWER 4–7 cm (1.5–2.75 in) long, terminal racemes. FRUIT 13–32 cm (5–12 in) long.

The Mexican name "mata ratón" (*matar* = to kill; *ratón* = rat) is derived from the poisonous qualities of the seeds and bark, which are pulverized and fed to rats or other rodents. Because it is a fast-growing tree, early settlers used it to shade cacao bean plantations, thus the Mexican common name "madre cacao" and the Nahuatl "cacahuananche," meaning cacao mother.

This species flourishes along both coasts of Mexico: on the east from Tamaulipas south to Quintana Roo and Chiapas, and on the west from Sinaloa to the west coast of Chiapas.

LEGUMINOSAE—PEA FAMILY

Gliricidia sepium

TREE to 12 m (40 ft) tall, gray bark. LEAVES pinnate; 7–17 leaflets 3–7 cm (1.75–3 in) long. FLOWER 2.5 cm (1 in) long, pink, racemes. FRUIT 15–20 cm (6–8 in) long, 2–3 cm (.75–1.25 in) wide.

Olneya tesota Gray
Palo Fierro or Ironwood

Olneya tesota, a spreading, thorny tree with fissured gray bark and pubescent branches, is armed with short, straight or curved spines. These thorns are brown or brown with black tips and are arranged singly or paired beneath the gray-green leaves, which are covered with fine

223

LEGUMINOSAE—PEA FAMILY

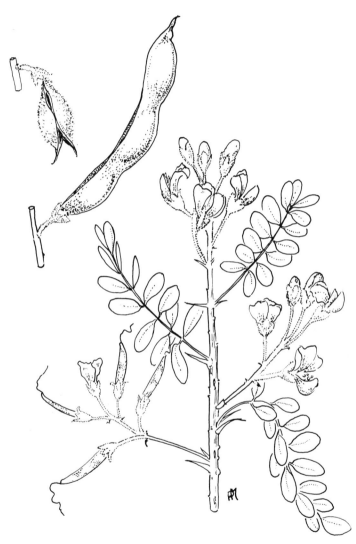

Olneya tesota
TREE 5–10 m (16–33 ft) tall, armed with spines 5–10 mm long. LEAVES pinnate 3–10 cm (1.25–4 in) long; leaflets 5–10 mm long. FLOWER 9–10 mm long; calyx 5–7 mm long. FRUIT 3–6 cm (1.25–2.5 in) long, 8–9 mm broad.

LEGUMINOSAE—PEA FAMILY

hairs. Pinnately compound, the leaves may develop an even or odd number of small leaflets. The few-flowered axillary inflorescences have the typical pealike flower: whitish, tinged with purple or maroon, resembling miniature wisteria blooms. Dehiscent seed pods are constricted on both sides of the one to five edible black seeds. The extremely hard, heavy, beautiful wood is used by the Seri Indians in their famous carvings.

Flowering in May and June, *Olneya* is common in desert washes and on low hills in Sonora and Baja California. It usually is an indicator of a nearly frost-free area, for its young branches are frost sensitive. Its presence is an indicator of suitable areas for citrus culture.

Sophora secundiflora (Ort.) DC.
Frijolito

One of the most attractive spring-flowering plants of northern Mexico is *Sophora secundiflora*, with its many pendant clusters of fragrant, lavender, wisterialike flowers. The specific name *secundiflora*, meaning side-flowering, may have been chosen because the pea-shaped blooms are densely crowded on one side of the axis of the raceme. Of the five unequal petals, the large, two-lobed upper petal (banner), standing erect with margins curling back, is twice as broad as the other petals. It is light lavender, striped with darker veins, and has a white mid-portion. The two dark purple wing petals enfold the 10 stamens and pistil and hide from view the two dark purple lower (keel) petals. The cylindrical, woody, slightly fuzzy seed pod, with a curled, stiff-pointed apex, is very slightly constricted between the seeds. These conspicuous, light yellowish-tan, hard pods, persisting up to a year, contain one to three vermillion or bright red, extremely poisonous seeds. Lustrous deep green leaves are pinnate, with an unpaired terminal leaflet to each leaf.

Being evergreen and abundantly foliated, this plant is especially sought as a cultivar. It grows natively from Coahuila to San Luis Potosí.

LEGUMINOSAE — PEA FAMILY

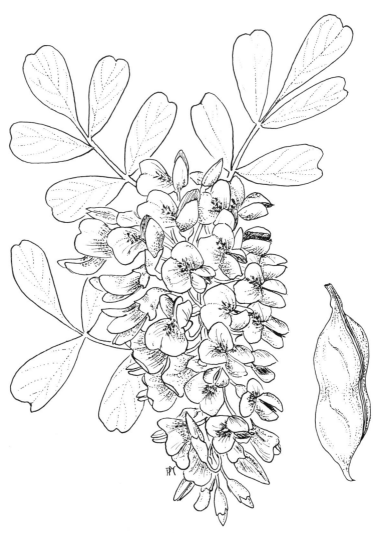

Sophora secundiflora

SHRUB or tree to 12 m (40 ft) tall, evergreen. LEAVES pinnate, 7–11 leaflets 2.5–6 cm (1–2.5 in) long, obovate. FLOWER 2–3 cm (1.75–2.25 in) long, lavender; raceme. FRUIT to 20 cm (8 in) long, 1.5 cm (.75 in) thick, woody, cylindrical.

Lobeliaceae — Lobelia Family

Lobelia laxiflora H.B.K.
Aretitos or Lobelia

A profusion of red flowers, borne on elongated pedicels in leafy racemes, transforms *Lobelia laxiflora* into a colorful plant from March through October. The cylindrical red corolla tube is five-lobed, opening into two lips; the lower three lobes are yellow, while the upper two are red on the outside and orange inside. The upper, lanceolate lobes stand erect or are slightly recurved. Projecting over the lower lip are the five large, white, fuzzy, united anthers, whose filaments are fused into a column surrounding the pistil. The fruit develops into a capsule with many small seeds and crowned with the short, five-toothed, green calyx. Acuminate, sessile or short-petiolate leaves have toothed margins and often are densely pubescent on the lower surface.

There are numerous "Lobelias" in Mexico, both annual and perennial, herbaceous and shrubby. *L. laxiflora* grows from Baja California and Sonora to Oaxaca and Veracruz.

Loganiaceae — Logania Family

Buddleia cordata H.B.K.
Tepozánn

Of the many species of *Buddleia* in Mexico, *B. cordata* can be recognized by its dense, highly branched, terminal panicles of small fragrant flowers. The minute, four-petaled, campanulate corolla may be white, cream, or yellowish with orange at the throat. Although there are many flowers to a panicle, each cluster consists of but a few short-pedunculate blossoms. Male and female flowers are borne on separate plants. Showy, light brown, small, cylindrical fruiting capsules, each with up to 100 minute winged seeds, replace the female blooms after pollination. Ovate to lanceolate opposite leaves on relatively short petioles are dark green and essentially glabrous above, but whitish or yellowish beneath as a result of a dense coating of stellate hairs.

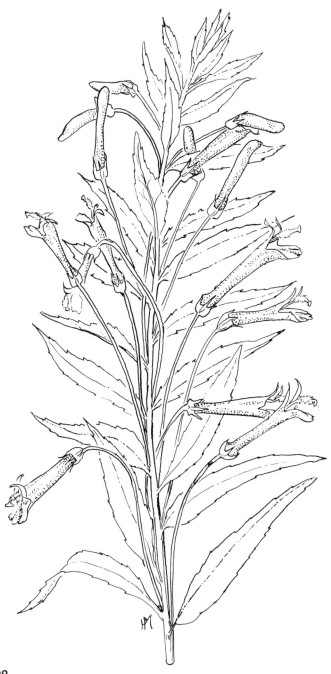

Buddleia cordata

SHRUB or tree 2−12 m (6.5−40 ft) tall, dioecious. LEAVES 4−23 cm (1.5−9 in) long, 3−14 cm (1.25−5.5 in) wide, ovate to lanceolate. FLOWER 1.5−2.5 mm long, clustered in terminal panicle 6−30 cm (2.5−12 in) long. FRUIT 3.5−5 mm long, capsule; seeds numerous.

Lobelia laxiflora

HERB or shrub to 1 m (3 ft) tall. LEAVES 6−20 cm (2.5−8 in) long, 1−5 cm (.5−2 in) wide, sessile. FLOWER 3−4 cm (1.25−1.5 in) long, red, cylindrical. FRUIT a capsule.

This species of *Buddleia*, which is widely distributed and common in the pine and oak forests of the highlands, may be found from Chihuahua, Coahuila, Nuevo Leon, and Tamaulipas south to Chiapas.

Loranthaceae — Mistletoe Family

Psitticanthus calyculatus (DC.) D. Don
Ingerto

Brilliant red flowers and fleshy green leaves growing in conjunction with leaves of a different shape and texture signal the parasitic shrub *Psitticanthus*, which grows on a variety of hosts. The large bright red or red-orange flowers have six linear petals, to each of which a prominent stamen is partially united. The sepals are present as a small cup at the base of the petals. In bud the petals are curved and thickened at the apex. Oval black or purple-black fruits are juicy and sticky. Fleshy, opposite leaves on coarse quadrangular stems are large, leathery, and variable in shape, but usually lanceolate, curved, or oblique on short petioles.

Most commonly the host plant is a broad-leafed tree, but occasionally this colorful parasite can be found growing on needle-bearing conifers. Unfortunately, "ingerto" usually grows high in the trees, making it difficult to study. It occurs from Tamaulipas south in the warmer parts of Mexico, and into Guatemala and Central America.

A peculiar, swollen, flowerlike growth remains when *Psitticanthus* is broken from the host plant. Called "flor de madera" or "flora de palo," the highly prized structures are gathered for sale in market places to become decorations in the home.

Lythraceae — Loosestrife Family

Cuphea jorullensis H.B.K.
Hierba del Cáncer

Although a small, perennial, herbaceous plant, *Cuphea jorullensis* makes a colorful display when growing in masses. The red-orange or

LORANTHACEAE—MISTLETOE FAMILY

Psitticanthus calyculatus

SHRUB 1 m (3 ft) tall, parasitic. LEAVES 6–15 cm (2.5–6 in) long, glabrous, on angulated stems. FLOWER 3–5 cm (1.25–2 in) long, red or red-orange. FRUIT 1 cm (.5 in) in diameter, globose, black berry.

red flowers, each arising on a prominent pedicel from the axil of an upper leaf, form small clusters at the apex of the stem. An inflated, calyxlike, tubular structure (hypanthium) grades from red-orange to green. At the lower end it forms a small spurlike sac, and at the upper end terminates in six small sepals and six minute white petals. Ten or eleven stamens protrude beyond the end of this floral cup. Sandpaper-rough (scabrous) hairs cover the stem, pedicels, leaf surfaces, and hypanthium. The lanceolate leaves on short petioles are opposite in arrangement. The fruit, a small papery capsule, produces 10 to 20 seeds.

Cuphea becomes abundant in pastures and openings in pine forests from Durango to Oaxaca. The epithet *jorullensis* is derived from the extinct volcano Jorullo in Michoacán, where the species was first found.

Magnoliaceae — Magnolia Family

Magnolia schiedeana Schl.
Corpus or Palo de Cacique

Magnolia schiedeana is a stately, handsome evergreen even when not in bloom, by virtue of its large lustrous leathery leaves. When blossoming, the tree is striking; its showy, creamy-white, open flowers are borne solitarily in axils of the leaves or at the ends of the branches. Six large, obovoid, leathery petals are joined by the three white sepals to form a chalicelike blossom. Countless stamens and pistils in spiral formation create a conspicuous, central, erect, columnar mass. An interesting conelike fruit develops, and at maturity many vivid red seeds hang by thin threads from the fruit.

Endemic to Mexico, *M. schiedeana* grows on the Pacific slopes from Sonora, Chihuahua, Durango, Sinaloa, and Nayarit to Veracruz on the Atlantic slopes. Another tree of the Magnolia Family, similar but more southerly in distribution, is *Talauma mexicana* (DC.) Don, which with its white, showy, fragrant flowers resembles *M. schiedeana*, but differs in that its fruits do not open to disperse the seeds. Both species were revered by Indian royalty, as the common name "palo de cacique" (tree of the Indian chief) denotes. *Talauma* grows from Puebla and Veracruz to Chiapas, and on the Pacific slope from Guerrero to Guatemala.

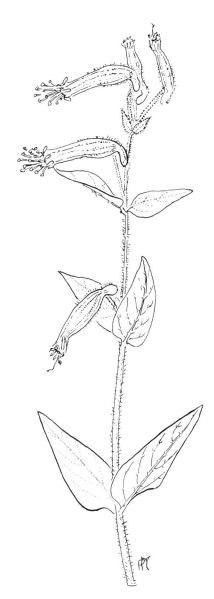

Cuphea jorullensis
HERB to 80 cm (32 in) tall, perennial. LEAVES 2–6 cm (.75–2.5 in) long, lanceolate, scabrous; petiole 1–5 mm long. FLOWER 20–28 mm long; petals minute, white. FRUIT capsule with 10–20 seeds.

MAGNOLIACEAE — MAGNOLIA FAMILY

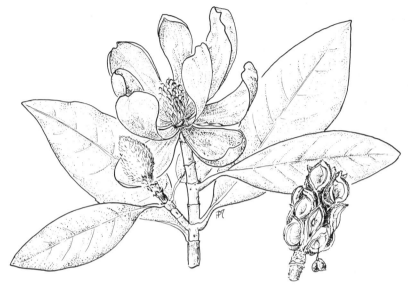

Magnolia schiedeana
TREE 3-25 m (10-80 ft), evergreen. LEAVES 4.7-16 cm (2-6.25 in) long, 2.5-7.7 cm (1-3 in) wide, elliptic. FLOWER 12-20 cm (5-8 in) in diameter, solitary, white. FRUIT 4-8 cm (1.5-3 in) long, conelike; seeds 5-7 mm long, red.

Malpighiaceae — Malpighia Family

Byrsonima crassifolia (L.) H.B.K.
Nanche

Byrsonima is a small, crooked tree or shrub of the grasslands, savannas, and dry forests. Its opposite, elliptic, thick, evergreen leaves are covered with fine rusty hairs when young, becoming glabrous with age (*crassifolia* = thick-leafed). Showy yellow flowers, borne in terminal raceme clusters, have pale yellow-green sepals, five bright yellow-orange petals, each narrowed into a stalklike base (claw), and 10 conspicuous, erect stamens. Round drupaceous fruits contain one hard stone surrounded by a sour, but edible, juicy pulp, which is eaten raw or cooked. The fruits usually are available in the market.

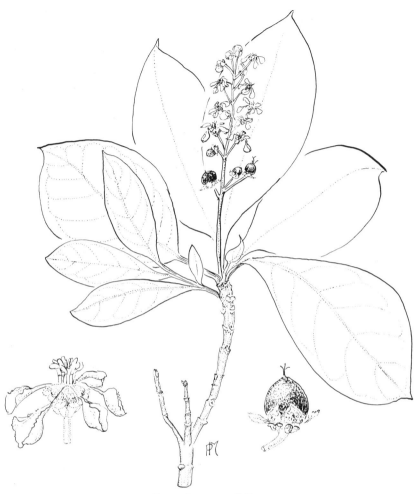

Byrsonima crassifolia

SHRUB or tree to 7.5 m (25 ft) tall. LEAVES 4–15 cm (1.5–6 in) long, opposite, elliptic. FLOWER 2 cm (.75 in) wide, in racemes 7.5–10 cm (3–4 in) long; petals yellow-orange. FRUIT 1 cm (.5 in) in diameter, yellow, globose.

This plant is found in the warm grassland areas from Sinaloa south to Chiapas and central America and east to Veracruz and the Yucatán Peninsula. It also grows in some of the warm oak forests of Veracruz, Oaxaca, and Tabasco.

MALPIGHIACEAE—MALPIGHIA FAMILY

Galphimia glauca Cav.
Lluvia de Oro

The graceful racemes of bright yellow flowers that cover *Galphimia* inspired the common name "rain of gold" (lluvia de oro). The numerous starlike blooms are composed of five regular, ovate, yellow petals, each tapering to a conspicuously narrow base referred to as a claw. Ten scarlet filaments surround the pistil, which develops into a small, three-lobed, oval fruit. Glossy, green, opposite leaves are on petioles almost as long as the ovate blades, at the base of which are two small glands. The young branches and the axes of the racemes of *Galphimia* are reddish.

"Lluvia de oro," which flowers most of the year, is found from Sonora east to Tamaulipas, and south to Chiapas into Central America. It has occasionally been used as a showy cultivated plant.

Mascagnia macroptera (Moc. & Sessé) Niedenzu
Gallinita

Clusters of yellow flowers on a low, scrambling, trailing vine, or subsequent groups of light-colored, winged fruits, mark *Mascagnia*. A closer look at the appealing flower reveals five small sepals with eight to ten reddish-purple to black glands, which at times are so large and conspicuous as to almost hide the calyx lobes; on other specimens they are lost among the hairs that clothe the surface. Five separate, ovate to orbicular petals narrow into a short but conspicuous claw. Within the flower are 10 stamens and a three-lobed ovary that eventually develops into a fruit with prominent wings, which aid in wind dispersal of the seed (*macro* = large; *ptera* = wing). Opposite, ovate to lanceolate leaves on short petioles complete the structure of the plant.

Usually coming into flower after rains, *Mascagnia* is most abundant in the northern states from Baja California to Tamaulipas, but it has been collected in several areas in Hidalgo and Veracruz.

MALPIGHIACEAE—MALPIGHIA FAMILY

Galphimia glauca
SHRUB to 5 m (16 ft) tall; reddish branches. LEAVES 6–8 cm (2.5–3 in) long, opposite. FLOWER 2.5–3.5 cm (1–1.5 in) broad, yellow, in racemes. FRUIT 6 mm in diameter, 3-lobed.

MALPIGHIACEAE—MALPIGHIA FAMILY

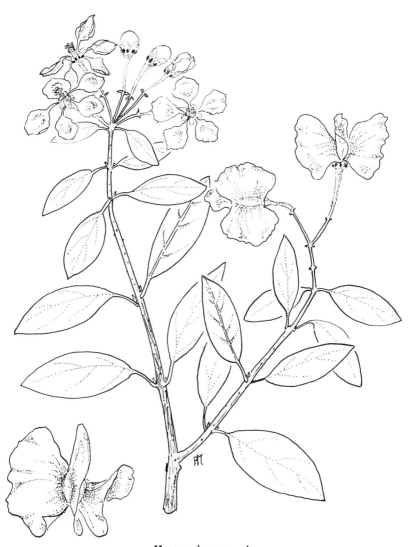

Mascagnia macroptera
SCRAMBLING VINE or shrub 2 m (6.5 ft) tall. LEAVES 2.5–8 cm (1–3 in) long, lanceolate to oval; petiole 2–3 mm long. FLOWER 2.5 cm (1 in) wide, yellow; sepals with glands. FRUIT 4–5.5 cm (1.2–2 in) broad, winged.

Malvaceae — Mallow Family

Bakeridesia notolophium (Gray) Hochreutiner
Mojagua or Canastilla

Bright yellow flowers on a shrub or small tree with large, attractive, heart-shaped leaves may be *Bakeridesia*, a relative of the hibiscus. Each flower has a densely hairy, tubular calyx and five spreading yellow petals. Standing erect from this open corolla is the conspicuous staminal column, which is formed by the fusing of the filaments of 250 to 300 stamens. The hibiscuslike flower arises singly or in pairs, more or less terminal, in the axils of reduced leaves. The leaves, deeply heart-shaped at the base, are dark green above but light green below, owing to a dense covering of star-shaped hairs.

This showy shrub grows in eastern San Luis Potosí and northern Veracruz. In many reference books this species is listed under *Abutilon*.

Martyniaceae — Martynia Family

Proboscidea fragrans Dcne.
Uña del Diablo or Devil's Claw

The strange horned fruits of this coarse glandular herb cannot easily be mistaken for any plant other than a *Proboscidea*. Large, showy, fragrant flowers are borne in racemes. The united, hairy sepals form a short tube with five obtuse lobes. A tubular, irregular corolla, deep violet or reddish-purple, opens with five spreading lobes exposing four stamens of two different lengths. After pollination, the style elongates and curls into a glandular, flesh-covered horn. As the fruit matures, the fleshy outer coat is shed and the horn splits into two stiff, curved hooks about twice as long as the body of the fruit. Broad deltoid leaves are covered with glandular pubescence.

This *Proboscidea* grows from the northern states of Chihuahua and Coahuila south to Puebla and Morelos.

MALVACEAE—MALLOW FAMILY

Bakeridesia notolophium
SHRUB or tree 3–5 m (10–16 ft) tall. LEAVES to 36 cm (14 in) long, cordate.
FLOWER 30–50 mm long, yellow; numerous stamens fused into a central column.

MARTYNIACEAE — MARTYNIA FAMILY

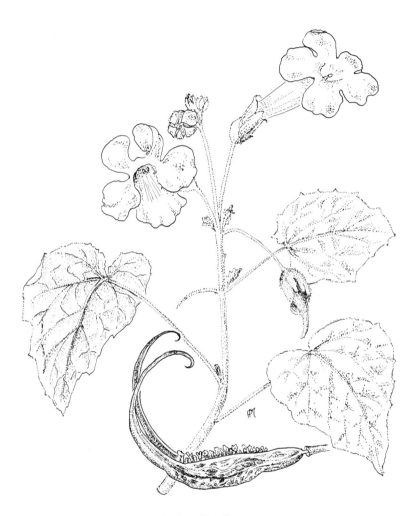

Proboscidea fragrans

HERB 50–60 cm (20–24 in) tall, annual, glandular pubescent. LEAVES 12 cm (5 in) long, 15 cm (6 in) wide, broad deltoid to ovate, 5-lobed; petiole 18 cm (7 in) long. FLOWER 5 cm (2 in) wide and long, fragrant, purple; in racemes of 10–20 flowers. FRUIT 30 cm (12 in) long; body 10 cm (4 in) long; horns 20 cm (8 in) long.

Melastomataceae — Melastome Family

Conostegia xalapensis (Bonpl.) D. Don
Capulincillo

The unusual contrast in color and texture between the upper and lower surfaces of the leaves makes *Conostegia* a conspicuous shrub or small tree. The opposite, ovate to oblong-lanceolate leaves are glabrous and a rich, dark forest green on the upper surface; a dense, felty tomentum covers and colors the under side with either whitish or pale brownish hairs. The leaves also are noticeable because of the five prominent veins spreading upward and outward from or near the acute or rounded base to converge again at the acuminate apex. These veins are, in turn, connected by smaller horizontal veinlets, looking like rungs of a ladder. Prominent dentate margins on the leaf add to the ease of recognition of this plant.

The spreading branches of the open paniculate inflorescence end in clusters of small pink flowers with 10 prominent stamens, or, at a later developmental stage, with dark blue or purple berries. The fruits, frequently available in the markets, are reminiscent of the northern blueberry in size, taste, and persistence of the calyx lobes.

From Sinaloa and Tamaulipas south to Veracruz, Tabasco, and Chiapas, *Conostegia* is common in moist or wet, open, bushy hillsides. It becomes abundant as second growth in cut-over areas, along fencerows, and in pastures, where seeds probably are spread by birds and cattle. Because the original collection of this plant was made in Xalapa, Veracruz, that city was honored in the specific name *xalapensis*.

Meliaceae — Mahogany Family

Cedrela odorata L.
Cedro or Spanish Cedar

"Cedro" is a large tree with a smooth gray bark when young, which becomes brown and furrowed with age. The evenly pinnate leaves have 10 to 20 pairs of ovate leaflets and become deciduous after the

MELIACEAE—MAHOGANY FAMILY

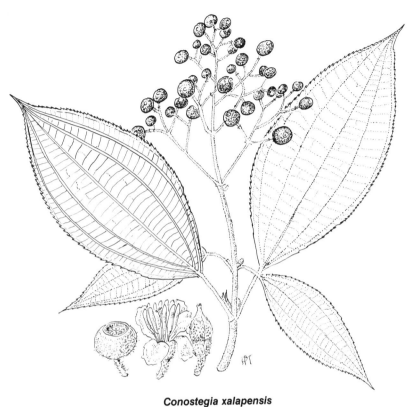

Conostegia xalapensis

SHRUB or tree to 6 m (20 ft) tall. LEAVES 8−20 cm (3−8 in) long, 2−7 cm (.75−3 in) broad, ovate to oblong-lanceolate; petiole 1−4 cm (.5−1.5 in) long, stout. FLOWER buds 5−7 mm long, petals 4−6 mm long, pink; anthers 2.5−3 mm long; panicle. FRUIT small, dark blue or purple berry.

fruit matures. From May to August large terminal panicles bear many small yellow-green flowers. More conspicuous than the minute flowers are the brown, woody, oblong-elliptic seed capsules. Pale lenticels give the fruit a warty appearance. The pods, containing many winged seeds, split open along the four or five valve sutures.

Cedrela grows in wet tropical areas along both coasts from Tamaulipas to Yucatán and Chiapas, and from Sinaloa to Guerrero.

MELIACEAE—MAHOGANY FAMILY

Cedrela odorata

TREE to 30 m (100 ft) tall. LEAVES 15–60 cm (6–24 in) long, evenly pinnate. FLOWER 7–10 mm in terminal panicles to 40 cm (16 in) long. FRUIT 2.5–5 cm (1–2 in) long, brown; winged seeds 20 mm long.

Although it flowers in summer, the fruits remain until winter. The name "cedro" comes from the cedarlike fragrance (*odorata* = fragrant) of the wood, which makes the lumber valuable for the manufacture of fragrant cigar boxes. The red-brown wood, very similar to mahogany, is easily worked and is highly prized for furniture. Because of the demand for the lumber, the tree has been overharvested. It grows rapidly and frequently is planted as an ornamental, but no significant attempt has been made to reforest cut-over areas.

Swietenia humilis Zucc.
Caoba

Large seed capsules standing upright on a tree like candles, will attract attention to *Swietenia humilis*, a mahogany tree. The noticeable fruit is ovoid, brownish-gray, woody, and stands erect with its pointed tip upward. To release its mass of glossy brown seeds, the five boat-shaped segments split from the base, opening like an umbrella, and drop to the ground, leaving the winged seeds attached to the core to be loosened and scattered by the wind. The flower responsible for this large capsule is an amazingly small, five-petalled bloom that grows profusely on long-stalked axillary panicles. Centered on the spreading petals is a small cup-shaped disk made up of filaments of the 10 united stamens that surround the pistil. The pinnately compound leaves have ovate-lanceolate leaflets, whose asymmetrical halves join in tapering to a long, almost threadlike apex.

Swietenia humilis is one of the sources of the famed red-brown mahogany wood; the common name "caoba" means mahogany. It grows in the tropical regions from Michoacán to Chiapas.

Moraceae — Mulberry Family

Brosimum alicastrum Swartz
Ramón, Mojo, or Breadnut

Brosimum is a large evergreen tree with gray bark and smooth, bright green, elliptic leaves. The leaf apex ends in an abrupt, short point. In the fall, inconspicuous unisexual flowers bloom in rounded

Swietenia humilis

TREE to 10 m (32 ft) tall, wood red. LEAVES pinnate; leaflets 2–5 pairs 6–15 cm (2.5–6 in) long, lanceolate, long tapering tips. FLOWER 10–12 mm in diameter, white. FRUIT 15–20 cm (6–8 in) long, 10–12 cm (4–5 in) broad, capsule; seeds 6–9 cm (2.5–3.5 in) long.

MORACEAE — MULBERRY FAMILY

clusters with numerous staminate and one or two pistillate flowers combined in each head. The subglobose, yellow or orange fruit contains one large seed. The pulp of the fruit is edible, as is the seed after it is boiled.

Abundant milky latex present in this tree and common to many members of the Mulberry Family can be diluted with water and used

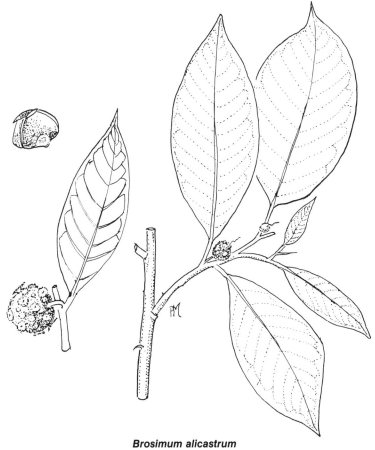

Brosimum alicastrum
TREE 30 m (100 ft) tall, evergreen, bark gray. LEAVES 7–14 cm (3–5 in) long, 3–5 cm (1.25–2 in) broad, elliptic. FLOWER small, unisexual. FRUIT 1.5 cm (.75 in) in diameter, yellow or orange.

as a substitute for cow's milk. *Brosimum* leaves are used for fodder. The wood is suitable for carpentry work, as it is light-colored, hard, and fine-grained. "Ramón," or "breadnut," is found in warm, moist tropical areas from southern Tamaulipas and Sinaloa south to Yucatán, Chiapas, Oaxaca, and into Central America.

Cecropia obtusifolia Bertol.
Chancarro or Trompeta

This member of the Mulberry Family probably is most easily recognized by its long-petioled, large, deeply palmately lobed leaves, which are green on the upper surface and conspicuously white or silvery underneath. *Cecropia* is a medium-sized tree with a widely spreading crown. The smooth, gray-barked trunk is solid, but the branches are hollow and frequently are inhabited by ants. Extremely large, alternate, orbiculate leaves on long, stout petioles are clustered at the ends of the branches. The unisexual flowers borne on separate trees are small but appear in great masses on long cylindrical spikes. Each tree, which flowers almost all the year, bears only one type of flower, either male or female.

Cecropia, widely distributed throughout the tropical countries, is found along the west coast from Sinaloa south to Oaxaca and Chiapas, and on the east coast from Tamaulipas to the Yucatán Peninsula.

FICUS SPECIES

A palm tree with a crown of unmistakably unpalmlike leaves, a mass of peculiar roots that seem to flow over and obscure the rocks of a canyon wall, a large tree with many pendant aerial roots trailing from the branches—all are introductions to the Mexican figs.

The genus *Ficus*, represented by about 17 species in Mexico, encompasses large and small trees with simple, entire leaves. Insignificant flowers are so small that descriptions are omitted in the entries that follow. The blossoms are borne on the inner surface of a more or less globose receptacle, which has a small pore at the apex to admit pollinators, usually wasps. The fruits of all species are edible, but the

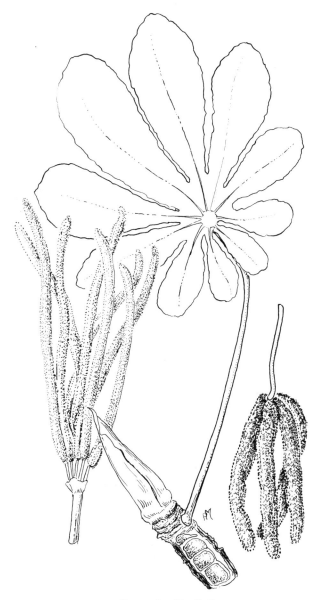

Cecropia obtusifolia
TREE to 14 m (45 ft) tall. LEAVES 30–75 cm (12–30 in) long, palmately lobed, green above, white below. FLOWER inconspicuous, unisexual, in spikes.

receptacles frequently are so small and dry as to be unpalatable to people, although they do furnish food for birds and other animals. *Ficus,* like many members of the family Moraceae, is distinguished by a milky sap, which often contains a type of rubber that in the future may have commercial value.

The common name for many species of *Ficus* is "amate," a Nahuatl name meaning paper; before the Spanish conquest, the Indians used fig bark to make paper. The bark was stripped from the tree, soaked in lye water, washed, boiled, and split into thin strips. These were then placed on a plank and pounded with a stone until a sheet of paper resulted, a process not greatly different from the production of papyrus paper by the Egyptians.

Unlike the common edible European fig, *F. carica*, which has lobed leaves and is widely cultivated in Mexico, native figs have simple leaves. Descriptions of all the figs is impractical here, but a few are outstanding and deserve recognition.

Ficus cotinifolia H.B.K.
Nacapuli

"Nacapuli" often is a large tree, bearing broadly oblong to nearly orbicular leaves with prominent veins on the lower surface. The small spotted fruits most frequently are sessile in axillary pairs.

This species, one of the strangler figs, may start its growth high above the ground in the crown of a palm tree, eventually extending its roots to the soil. The usual result is the death of the host plant. It is widely distributed throughout Mexico, reported from southern Sonora and southern Tamaulipas south to Oaxaca and Yucatán.

Ficus glabrata H.B.K.
Amate or Siranda

This species, with its pale, almost smooth bark, is one of the very large trees of southern Mexico. The crown is low, dense, and spreading; lanceolate to elliptic leaves have conspicuous lateral nerves. Thick peduncles support the solitary globose receptacles, which usually are

MORACEAE — MULBERRY FAMILY

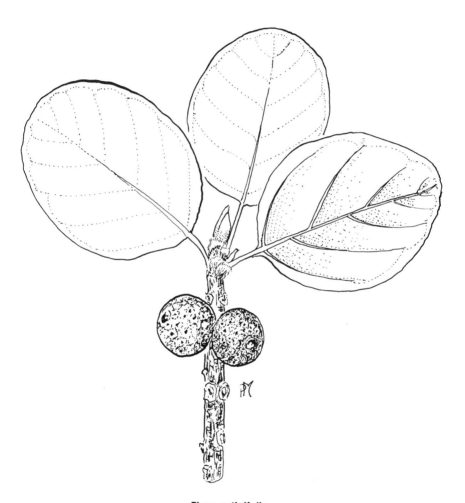

Ficus cotinifolia
STRANGLER TREE 15 m (50 ft) tall. LEAVES 5–14 cm (2–5 in) long, 2.5–10 cm (1–4 in) wide, oblong. FRUIT 6–11 mm in diameter, paired.

MORACEAE—MULBERRY FAMILY

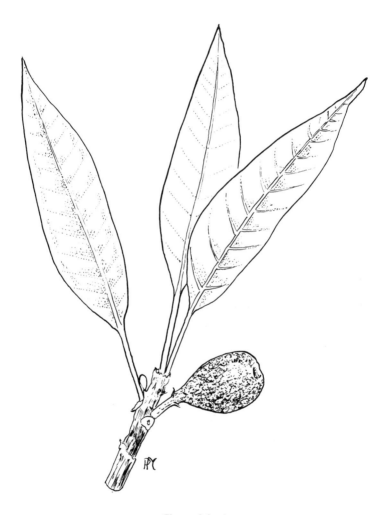

Ficus glabrata
TREE to 40 m (130 ft) tall. LEAVES 12–23 cm (5–9 in) long, 5–10 cm (2–4 in) wide. FRUIT soft, fleshy, about the size of a cultivated fig.

mottled light and dark green and become soft and juicy at maturity. The fruit is edible and approaches the size of the cultivated fig; although humans often find the flavor mediocre, the fruits are eagerly sought by other mammals and birds. Abundant latex issues from cut or broken stems. *Ficus glabrata* is found in Veracruz, Guerrero, Oaxaca and south into Central America.

Ficus padifolia H.B.K.
Matapalo or Strangler Fig

Ficus padifolia, a small or large tree with whitish or pale yellow bark, frequently starts as an epiphyte when a bird deposits a seed near the top of a palm tree. Its roots eventually reach the ground; as it continues to grow it envelops and strangles the host, hence the name "matapalo" (killer tree). When well established, this fig produces a broad, spreading crown, from which numerous aerial roots develop, giving the tree the appearance of a banyan. Leaves on long slender petioles are lanceolate to lance-oblong. Axillary green receptacles, paired at the nodes on short slender stalks, are subglobose, usually spotted with red or purple; children, especially, seek them to eat.

"Matapalo" is one of the most widely distributed, variable, and abundant of the strangler figs. It occurs from southern Sonora and southern Tamaulipas south to Veracruz, Tabasco, Oaxaca, and into Central America.

Ficus petiolaris H.B.K.
Amate

Ficus petiolaris is a small to large tree with whitish to yellowish bark and rounded, heart-shaped leaves on long slender petioles. The leaf veins arise like fingers from the end of the petiole, and a tuft of white hairs at the base of the principal veins is characteristic. Small globose receptacles are on peduncles in axillary pairs and become mottled as they mature. This species is common along the west coast from Sonora south to Guerrero, where it grows on stream banks and often on cliff faces, looking as if it had been poured from a rock crevice.

MORACEAE — MULBERRY FAMILY

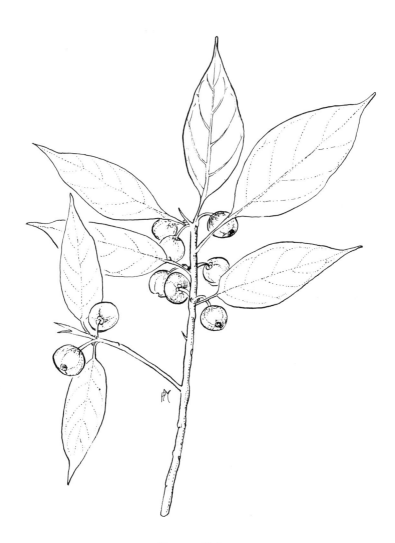

Ficus padifolia
TREE to 30 m (100 ft) tall. LEAVES lanceolate, 4–12 cm (1.5–5 in) long. FRUIT 9–12 mm in diameter, in axillary pairs.

MORACEAE — MULBERRY FAMILY

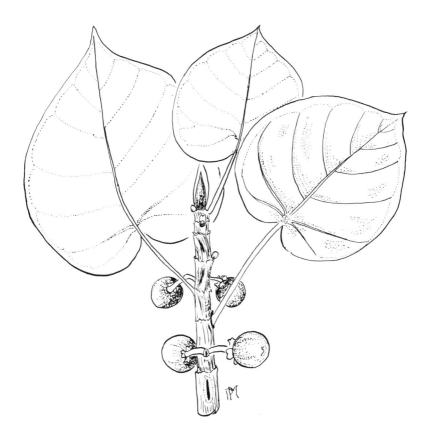

Ficus petiolaris
SMALL to large tree. LEAVES 6.5–15 cm (2.5–6 in) broad, cordate orbicular. FRUIT 10–15 mm in diameter, globose, paired.

Musaceae—Banana Family

Heliconia latispatha Benth.
Platanillo or Golden Heliconia

A spectacular and bizarre plant of the hot wet forests, *Heliconia latispatha* bears brilliantly colored inflorescences, which, unlike most species of *Heliconia*, rise erect above paddle-shaped green leaves. Pointed golden-orange bracts, the conspicuous part of this plant, remind one of *Strelitzia*, the bird-of-paradise flower. The bracts are narrowly lanceolate, spreading, and long-attenuate; the lowest one often dilates at the apex into green blades resembling leaves. Almost hidden among the bases of the long bracts are small, inconspicuous, pedicellate flowers with a greenish-yellow perianth. Long-petiolate leaves, often glaucous on the under surface, are oblong with short acuminate tips.

Heliconia latispatha (*latispatha* = broad spathe) is a relative of the banana, as its Mexican common name implies (*"platanillo"* = little banana), and as such flourishes in frost-free, humid areas in Veracruz, Chiapas, and the southernmost states of Mexico.

Nyctaginaceae—Four-o'clock Family

Bougainvillea glabra Choisy
Bugambilia

Commonly cultivated throughout the frost-free areas of the world, *Bougainvillea*, a Brazilian native, makes a brilliant show with its long sprays of purple-magenta to rose-red bracts. The plant is a spiny vine or sprawling flamboyant shrub, which may extend three meters or more if not trained or supported. Alternate, ovate, glossy green leaves subtend the masses of showy bracts, each with its single, five-angled, tubular,

---

Heliconia latispatha
HERB 1.5–2.5 m (5–7 ft) tall, glabrous, stout. LEAVES to 1 m (3 ft) long, 20–30 ▶ cm (8–12 in) wide, long-petiolate. INFLORESCENCE erect. FLOWER 3–3.5 cm (1.25–1.5 in) long, inconspicuous; bracts to 15 cm (6 in) long, 1.5–2 cm (.5–.75 in) high at base, orange-yellow.

MUSACEAE—BANANA FAMILY

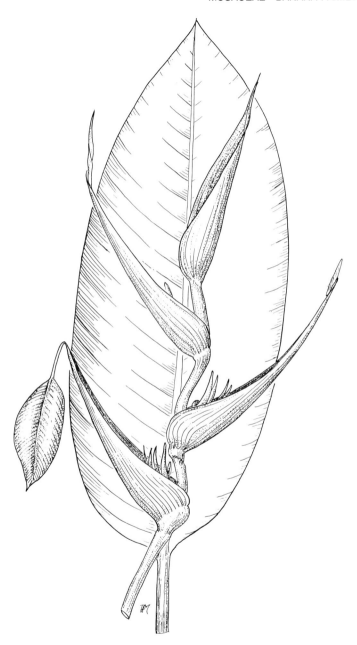

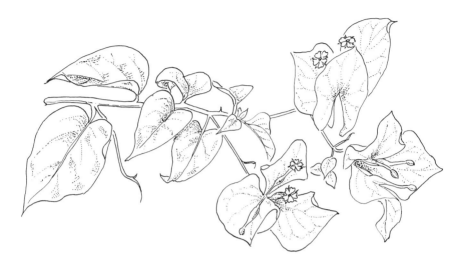

Bougainvillea glabra

VINE or sprawling shrub 3+ m (10+ ft) long, spiny. LEAVES 5–10 cm (2–4 in) long, broadly ovate, acute at apex. FLOWER insignificant; showy bracts 2–5 cm (1–2 in) long. FRUIT 11–14 mm long, grayish-green.

insignificant, cream-colored flower. The heart-shaped bracts come in a wide range of colors, from the most common purple-magenta to red, orange, pink, golden yellow, and white.

Bougainvillea, found throughout Mexico—trained on walls, growing over rooftops, or even climbing in trees—was named for Louis de Bougainville, the eighteenth-century French navigator who discovered this plant in Rio de Janeiro.

Pisonia capitata (S. Wats.) Standl.
Garambullo

This densely branched shrub or small tree spreads its grayish stems or branchlets at right angles. The clambering branches may extend to great lengths if supported by other vegetation. Stout, straight or curved, cat-clawlike spines 7 to 14 mm long are produced in axils of the orbicular or obovate leaves. Young growth, leaves, and inflorescences are finely

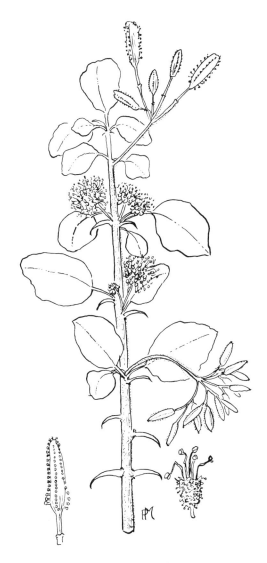

Pisonia capitata

SHRUB or tree to 5 m (16 ft) tall, spiny. LEAVES 2–6 cm (1–2.5 in) long, opposite, on long petioles. FLOWER unisexual, small. FRUIT 10 mm long with stalked glands.

hairy. *Pisonia* flowers are unisexual; the plants usually are dioecious, although monoecious individuals do occur. Small, fragrant, red male flowers clustered in dense masses (*capitata* = in heads) produce at maturity five conspicuous, yellow, exserted stamens. The tubular female flowers are arranged in loose, open panicle-type inflorescences. Each produces a brown prismatic fruit that bears a row of sticky glands on its five angles.

This species, which blooms in early spring, grows in Sonora, Sinaloa, Durango, and Nayarit. A similar species, but one with more yellowish male flowers and smooth shiny leaves, *P. aculeata* is eastern and more widely distributed. It grows from Tamaulipas south to Veracruz, Oaxaca, and into Central America; some citings are from Nayarit and Jalisco southward in western Mexico.

Onagraceae—Evening Primrose Family

Fuchsia fulgens DC.
Flor de Arete

Fuchsia is a low ornamental shrub with firecrackerlike flowers freely hanging in short, terminal, leafy racemes. The long, thin, pendant floral tube, which develops from a green inferior ovary, gradually widens at the mouth. At this rim of the tube are four reddish ovate-lanceolate sepals tipped with yellow-green or green, and four deep scarlet oblong-ovate petals about half as long as the sepals. Protruding beyond the sepals are eight stamens and a longer pink style ending in a round green stigma. After pollination, the pendulous, withered floral tube falls from the ovary, which develops into a dark green or purplish green, ellipsoid, fleshy, edible fruit.

Soft woody stems arise from a tuberous-thickened root. The young, somewhat succulent branches and leaves are finely pubescent and tinged with red. Opposite, broadly-ovate to oblong-ovate leaves with rounded or heart-shaped bases and finely serrated margins have a reddish central vein.

A plant of rocky places, woodlands, and open areas, this species may occasionally grow as an epiphyte. It is native to Jalisco, Michoacán, Chiapas, and Veracruz. The genus *Fuchsia* is widely distributed and is

ONAGRACEAE — EVENING PRIMROSE FAMILY

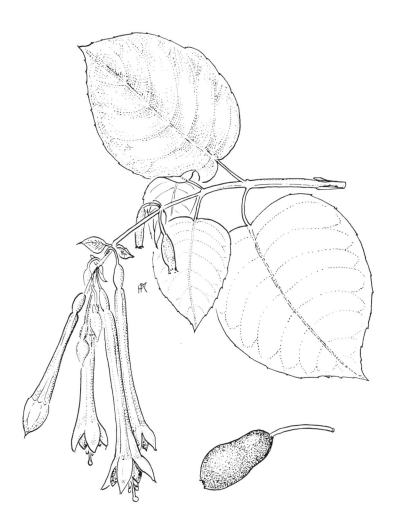

Fuchsia fulgens
SHRUB 30–120 cm (12–48 in) tall; stems soft, woody. LEAVES 5–17 cm (2–7 in) long, 3–12 cm (1.25–5 in) wide, ovate to oblong-ovate. FLOWER 5–7.5 cm (2–3 in) long, short racemes; sepals 12–14 mm long, ovate-lanceolate; petals 8 mm long, red. FRUIT 2+ cm (1+ in) long, ellipsoid.

popularly cultivated for its showy multicolored flowers. It was named to honor Leonhard Fuchs, an outstanding sixteenth-century German botanist.

Orchidaceae—Orchid Family

Mexico is well supplied with countless species of terrestrial and epiphytic orchids. The terrestrial forms usually are secreted in forested areas out of the way of the casual visitor. The epiphytic species are found growing on the rough bark of trees in areas of warmth and abundant moisture, commonly in the company of ferns, bromeliads, and other epiphytes. One of the many species represents this large and beautiful family here.

Epidendrum parkinsonianum Hook.
Pata de Paloma or Bayoneta

Epidendrum parkinsonianum, a widely distributed epiphyte, has fragrant, showy, white flowers borne in clusters of three at the apex of a dark green pseudobulb covered in papery scales. The fleshy, stiff, thick perianth parts are long-lasting, eventually turning in age to an attractive, mellow gold. The flower is composed of three linear lanceolate sepals, two broader lateral petals, and an enlarged three-lobed lower petal—a lip consisting of two half-circles flanking a linear central lobe. With a bit of imagination, the outline may be seen to resemble the footprint of a dove, giving rise to the common name "pata de paloma." One thick fleshy leaf, channeled its full length, develops at the apex of the pseudobulb.

This species grows on trees and rocks in pine-oak forests from Sinaloa to Oaxaca and Chiapas, then southward through Central America to Panama. It also is found in Puebla and Veracruz.

Epidendrum parkinsonianum
EPIPHYTE, pendant, herbaceous, to 2 m (6.5 ft) long. PSEUDOBULBS present, ▶
6–10 cm (2.5–4 in) long. LEAVES 20–50 cm (8–20 in) long, fleshy. FLOWER 3.5–5.5 cm (1.25–2 in) long, white; lip 3-lobed. FRUIT 7–13 cm (3–5 in) long, 3-ridged capsule.

ORCHIDACEAE — ORCHID FAMILY

Palmae—Palm Family

The palms form a large and complicated group of plants. Only the seven most likely to be seen are included here. One in particular, an introduced species, is mentioned because of its economic importance.

In general, the palms can be divided into two groups based upon the shape of the leaf: palmately lobed (fan-shaped), or pinnately compound (resembling a feather). Both types are represented in the following species.

The trunks or stems of the palms under consideration are unbranched, bear a tuft or crown of leaves at the top, and are found with

Palmately lobed (fan-shaped) and pinnately compound (featherlike) palm leaves.

or without a skirt of dry dead leaves below the green ones. The flowers in large panicles are small, either unisexual or bisexual (perfect). Fruits are variable in size from small globular structures less than 2.5 cm (1 in) in diameter to large coconuts up to 15 to 20 cm (6–8 in) wide. The stems and leaves are highly valued, as are the edible seeds of some and the fruit coat of others. Stems are used in some of the rural areas for construction material. Leaves, besides serving as thatching for homes or ramadas, are woven into mats, purses, hats, bedding, ornaments, toys, and art objects.

FEATHER PALMS

Acrocomia mexicana Karw. ex Mart.
Coyol

The featherlike leaves of *Acrocomia* are well protected by long sharp spines on the base, petiole, and rachis; the spiny bases of fallen leaves armor the stout trunk for an indefinite period. Both sexes of small, unisexual flowers bloom in the same long, dense, spiny panicle from March to July. Tightly clustered, globose fruits have a thin but hard outer covering surrounding a fibrous middle layer, with a very hard inner portion protecting the single seed. Only the edible, sweet, oily seed of this palm is used; these often are found in the markets.

Occurring on both the Atlantic and Pacific slopes, "coyol" ranges in the east from southern San Luis Potosí, northern Puebla, and Veracruz to Chiapas, Tabasco, and the Yucatán Peninsula. On the western slope it can be found from Sinaloa south to Chiapas.

Cocos nucifera L.
Palma de Coco or Coconut

Commonly a coastal tree, *Cocos* evokes images of sandy tropical seashores. Erect, or more commonly leaning as a result of constant ocean winds, it bears a tuft of large pinnately compound leaves. Dead leaves are persistent for only a short time; when they fall, a large, smooth, conspicuous scar is left. Numerous small unisexual flowers

PALMAE — PALM FAMILY

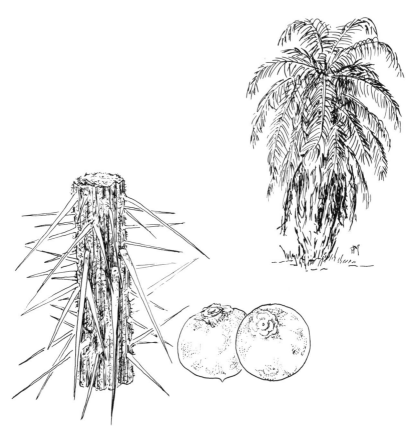

Acrocomia mexicana
TREE to 10 m (33 ft) tall; spiny, persistent leaf bases. LEAVES pinnate, 4 m (13 ft) long; leaflets 1 m (3 ft) long. FLOWER unisexual; panicle 80 cm (32 in) long. FRUIT 4 cm (1.5 in) in diameter, globose; seed 1.5 cm (.75 in) in diameter.

develop with both male and female flowers in the same panicle. The pistillate flower develops into the coconut, which is the second largest fruit of all the world's palms (*nucifera* = nut bearing).

Worldwide in its distribution, *Cocos* frequently is cultivated in the tropical regions along both coasts of Mexico. All parts of the plant are

PALMAE—PALM FAMILY

Cocos nucifera
TREE 25–30 m (80–100 ft) tall. LEAVES pinnate 3–6 m (10–20 ft) long; leaflets 75–100 cm (30–40 in) long. FLOWER small; panicle 1–1.5 m (3–4.5 ft) long. FRUIT (coconut) 25–30 cm (10–12 in) long, 15–20 cm (6–8 in) wide.

useful. The fruits are especially valuable, for the nut provides both food and drink and is the source of the coconut oil used in cosmetics and prepared foods. Even the fibers from the outer husk are woven into rugs and mats, while the hard inner shells may become cups and other eating utensils; the stem and leaves are used in construction and thatching.

Orbygna guacuyule (Lieb.) Hernandez X.
Guaycoyul

Orbygna is a tall tree with a crown of featherlike leaves subtended by a thin skirt of pendant dead leaves. An armor of leaf bases persists for some time after the dead leaves have fallen. Small, cream-colored, unisexual flowers are densely clustered in elongated panicles from March to May. An inflorescence may contain only male flowers, or it may consist of both male and female. The resulting nuts resemble small coconuts and hang on the tree for many months. Persistent perianth parts enlarge with the maturing fruit to envelop the lower one-third to one-half of the nut, while at the opposite end the style enlarges and becomes a conspicuous, apical point. Each nut contains one to three oily, edible seeds of commercial importance as a source of fats for the manufacture of vegetable oil. This handsome palm occurs on the Pacific slope from Nayarit to Oaxaca.

Scheelea liebmannii Becc.
Coyol Real

Scheelea is a common tall tree of undisturbed east coastal forested areas; it also grows in open, disturbed regions, where it may mature while relatively small. The large pinnately compound leaves of the forest-grown tall trees fall without producing a skirt, but the shorter trunks in open areas frequently are hidden by persistent dead leaves. Small, cream-colored, unisexual flowers are in dense panicles. Usually both sexes are in the same inflorescence, but in some individuals the number of male or female flowers may be greatly reduced, making the inflorescence almost unisexual. Flowering occurs from March to May. Hard, fibrous, dark or yellow-brown fruits, ovoid in shape and subtended by the persistent perianth segments, contain one to three oily seeds.

This palm forms groves, or palmares, along the Atlantic slope from northern Puebla and Veracruz to the northern part of Chiapas and Tabasco. As in most palms, the leaves may be used for thatching, but the most important products are the fruits, a valuable source of vegetable oil.

PALMAE — PALM FAMILY

Orbygna guacuyule
TREE to 30 m (100 ft) tall. LEAVES 5 m (15 ft) long, pinnate. FLOWERS unisexual; panicle 1.5 m (5 ft) long. FRUIT 6.5 cm (2.75 in) in diameter, ovoid or elliptic.

PALMAE — PALM FAMILY

Scheelea liebmannii

TREE to 30 m (100 ft) tall. LEAVES 7 m (25 ft) long, pinnate; leaflets 1.5 m (5 ft) long. FLOWER unisexual; panicle 1.3 m (4.5 ft) long. FRUIT 5–6 cm (2–2.5 in) long, ovoid.

FAN PALMS

Brahea dulcis (H.B.K.) Mart.
Palma de Sombrero, Palmito, or Cocaiste

Approximately seven species of *Brahea* have been reported for Mexico, but no concerted study has been made to delimit their distribution. They all have stiff, fan-shaped leaves and are distinguished on technical characters. One, a southern species, is discussed here.

The trunk of *B. dulcis* varies from erect to bent, or even prostrate. Under the spreading crown of stiffly erect fan-shaped leaves, dead leaves hang for a limited time as a skirt around the trunk. The flowers are perfect in elongated panicles that exceed the leaves in length. Small, yellow, ovoid fruits have only a thin edible flesh covering the seed. This species occurs mainly in Guerrero, Oaxaca, and Veracruz.

Sabal mexicana Mart.
Palma redonda or Palma Real

Fan-shaped leaves with a flexible blade are characteristic of *Sabal*. The curvature of the leaf is similar to that expected in a pinnately compound leaf with a short central axis. In age, instead of creating a dense skirt of dead leaves, only an area of persistent petiole bases may be apparent on the upper part of the trunk. Small, bisexual (perfect) flowers bloom in dense panicle inflorescences that approximate the length of the leaf. Dense, almost grapelike clusters of small, dark brown, globose fruits form on the branches or axis of the inflorescence, which persists long after the fruit has dropped.

This species of palm grows along the eastern coastal slope from southern Tamaulipas to the Yucatán Peninsula, and on the Pacific slope from Sinaloa to Chiapas. In addition to the common use of palm stems for construction and of leaves for thatching, it is reported that leaves of this species are used for making hats in Guatemala.

PALMAE—PALM FAMILY

Brahea dulcis
TREE to 7 m (23 ft) tall, erect or bent. LEAVES 1.5 m (5 ft) long, fan-shaped, stiff. FLOWER perfect; panicle exceeding leaves. FRUIT 1.6 cm (.75 in) long.

PALMAE — PALM FAMILY

Sabal mexicana
TREE to 20 m (65 ft) tall. LEAVES 2 m (6.5 ft) long, fan-shaped, flexible. FLOWER perfect; panicle 1.5 – 2 m (5 – 6.5 ft) long. FRUIT globose, 1.5 cm (.75 in) in diameter.

Washingtonia robusta Wendl.
Palma Blanca

Commonly used in ornamental planting, particularly in northwestern Mexico and southwestern United States, this tall stately palm at maturity develops a narrow head of fanlike leaves subtended by a long

Washingtonia robusta

TREE to 22 m (75 ft) tall. LEAVES fan-shaped blade 2 m (6.5 ft) wide and long; petiole 2 m (6.5 ft) long. FLOWER 8–9 mm long, perfect; panicle 3–4 m (10–13 ft) long. FRUIT 5–10 mm long.

PAPAVERACEAE—POPPY FAMILY

columnar skirt of dead leaves. In nature, the skirt may be variously interrupted where wind or other causes have loosened the dead leaves. Under cultivation, the skirt frequently is trimmed for a neat appearance or as a precaution against torching by vandals, leaving only the basal portion of the petiole, which develops a split herring-bone appearance. The petioles of leaves on young trees bear large hooked marginal spines; those on older trees tend to be smaller and less vicious. About 50 long segments form the broad fan blade; those in young trees are abundantly filamentous, while those on more mature plants are nearly naked. The leaves, except for the flexible tips, tend to be held in a horizontal plane. The split leaf base aids in distinguishing species of *Washingtonia* from those of *Erythea*, a genus of similar fan palms that may at times be used as ornamentals.

Numerous, whitish, bisexual flowers form inflorescences longer than the leaves. Exserted beyond three petals are six stamens with horizontal anthers. Large clusters of brown to black ellipsoid or ovoid fruits replace the flowers in the panicles. Each fruit is tipped with a persistent style and is subtended by three dry calyx segments. The naked panicle remains after the fruits have fallen and can be seen protruding from the skirt of dead leaves.

Washingtonia grows natively in canyons near coastal areas of Sonora, Baja California, and Sinaloa. The natural distribution may have been altered by Indians of the area, who ate the fruit and established new groves by discarding or planting the seeds.

Papaveraceae—Poppy Family

Argemone mexicana L.
Chicalote or Mexican Poppy

Large delicate flowers seem out of place among the prickly stems and spiny leaves of this showy, often weedy plant. The six separate, broadly obovate, thin petals often appear crinkled after they emerge from the three-horned bud, which is subtended by a conspicuous spiny bract. In the center of the yellow bowl-shaped flower a mass of bright orange stamens surrounds a prominent pistil capped with a feltlike, dark purple, lobed stigma. This showy stigma persists at the top of the seed capsule, which contains many small seeds and is well armed with stout spines interspersed with smaller ones. Spines also commonly develop on the stem, leaf margins, and lower surface midrib of the blue-gray, glaucous leaves, which are variable in size, shape, and lobing. Those near the base of the plant are oblanceolate; the middle and upper leaves are elliptical to ovate, occasionally clasping or sessile. All are lobed or toothed, often almost to the midrib; each tooth always terminates in a spine. A yellow or orange latex exudes when any part of the plant is injured.

Argemone usually grows as an annual, but under favorable conditions it may live for a few years as a perennial. It reacts as a weed, invading such disturbed areas as abandoned fields, arroyo embankments, and roadsides. Some botanists consider that two yellow species of *Argemone* exist in Mexico (*A. mexicana* and *A. ochroleuca*), but here they are treated as one, the differences being subtle. Although native to Mexico, this showy plant has now spread around the world and flourishes throughout the drier tropical and subtropical regions. In Mexico it is particularly common in the northern portion of the drier Central Plateau, but is reported from all states except the very southern ones.

Bocconia arborea S. Wats.
Llora Sangre or Palo del Diablo

Bocconia, a common roadside shrub, occasionally is tree-like. Large, pinnately lobed leaves, almost as broad as long, are clustered at the ends of branches. They are green on the upper surface and light green or gray on the lower surface. The few to many narrow lobes extend into tapering tips. In large terminal panicles are clusters of small flowers; as there are no petals, the greenish color is the outside of the two sepals. Two prominent, fuzzy stigmas appear before the sepals fall and expose the numerous stamens. At maturation of the black, dehiscent fruit, the valves fall away, leaving a ring of vascular tissue attached to the panicle branches.

PAPAVERACEAE — POPPY FAMILY

Argemone mexicana

HERB, annual or short-lived perennial 30–100 cm (12–40 in) tall; yellow or orange latex. LEAVES 35+ cm (14+ in) long, oblanceolate, spiny, irregularly and deeply lobed, glaucous. FLOWER 3–7 cm (1.25–3 in) in diameter, pale to dark yellow; stamens 20–75; pistil dark purple. FRUIT 20–50 mm long, capsule, spiny.

PAPAVERACEAE — POPPY FAMILY

Bocconia arborea tends to be western in distribution, from Sinaloa and Durango south to Oaxaca and Central America. As do many members of the Poppy Family, it has a yellow or orange sap ("llora sangre" = weeping blood), which may contain several alkaloids; consequently, these plants have medicinal significance. A yellow dye is obtained from the bark.

Bocconia arborea

SHRUB or tree 3–8 m (10–25 ft) tall. LEAVES 10–45 cm (4–18 in) long, deeply pinnately lobed, green above, light green or gray below. FLOWER small in terminal panicle. FRUIT 6 mm long, small black capsule.

Pinaceae—Pine Family

Two sources of characters are significant in the recognition of species of *Pinus*: the needles (leaves) and the cones. All needles are produced in fascicles, or clusters, and show a variation in length within a range; therefore, the number per fascicle and the overall length are important. Each fascicle is at first subtended by a sheath of scales. In some species the sheath is persistent throughout the life of the needles; in others it is deciduous and may fall immediately or may delay until the second year. The cones, which are the seed-producing structures, vary in size, shape, and longevity on the tree. To aid in the identification of the common pines, the following key is provided; it should be read in the same way as the key on page 11.

KEY TO THE COMMON PINES

1	Sheath at base of leaf fascicle deciduous	2
1	Sheath at base of leaf fascicle persistent	6
2(1)	Needles normally 5 per fascicle	3
2	Needles normally 3 per fascicle	4
3(2)	Cones globose, 6 cm (2.5 in) long	*P. leiophylla*
3	Cones elongated 20–45 cm (8–18 in) long	*P. ayacahuite*
4(2)	Needles 20–30 cm (8–12 in) long, drooping	*P. lumholtzii*
4	Needles less than 15 cm (6 in) long, not drooping	5
5(4)	Needles 2–5 cm (1–2 in) long	*P. cembroides*
5	Needles 6–14 cm (2.5–5.5 in) long	*P. chihuahuana*
6(1)	Needles 5 per fascicle	7
6	Needles 3 per fascicle	8
7(6)	Cones 8.5–15 cm (3.5–6 in) long	*P. montezumae*
7	Cones 5.5–8.5 cm (2–3.5 in) long	*P. oocarpa*
8(6)	Needles pendant, 15–30 cm (6–12 in) long	*P. patula*
8	Needles not pendant, less than 20 cm (8 in) long	*P. teocote*

* * *

PINACEAE—PINE FAMILY

Pinus ayacahuite Ehrenb.

Pino Cahuite or Acalo Cahuite

Elongated, pendant cones on stout peduncles are produced by this widely distributed large pine. The smooth grayish bark on young trees turns to a reddish-brown and becomes divided into irregular platelike scales as the trees age. Fascicles of five thin green needles with deciduous sheaths are characteristic of this species; the internal faces of the needles are markedly glaucous. Seeds show considerable variation in overall size, especially in wing size, which ranges from large in southern plants to nearly absent in northern trees.

This species occurs from the northern states of Sonora, Chihuahua, Coahuila, and Nuevo Leon southward to Veracruz, Oaxaca, and Chiapas, continuing southward into Guatemala.

Pinus cembroides Zucc.

Piñon

An important source of pine nuts (seeds) in Mexico, *Pinus cembroides* is one of the smaller pines, with a rounded top and many low branches resulting in a short trunk. An ashen bark is cracked and divided into thin, irregular plates. The fascicles consist most commonly of three needles, but as few as two or as many as four may occur. Variable also is the color of the needles from dark green to pale bluish green or yellow-green. Each cluster is subtended during its first year by a rosette or reflexed sheath scales, which are deciduous the second year. Small, early deciduous, globose cones are attached by short peduncles. The large, wingless, edible seeds with thick, hard shells are eagerly sought by man and animals for food.

"Piñon" grows on dry open slopes from the border states of Sonora, Chihuahua, Coahuila, Nuevo Leon, and Tamaulipas south through the dry central areas to Puebla, México, and Veracruz.

PINACEAE—PINE FAMILY

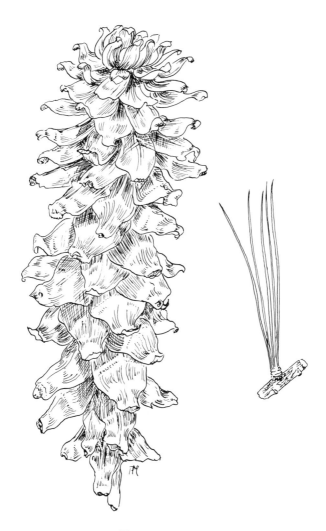

Pinus ayacahuite
TREE 20–30 m (65–100 ft) tall. LEAVES 8–15 cm (3–6 in) long, 5 per fascicle. SHEATH deciduous. CONE 20–45 cm (8–18 in) long, elongated, scale tips variously reflexed.

PINACEAE—PINE FAMILY

Pinus cembroides

TREE 5–12 m (15–40 ft) tall. LEAVES 2.5–5 cm (1–2 in) long, 3 per fascicle. SHEATH reflexed, forming rosette first year, deciduous second year. CONE small, 2.5–4 cm (1–1.5 in) long, 3–5.5 cm (1.25–2 in) wide; seed 15 mm long, edible.

PINACEAE – PINE FAMILY

Pinus chihuahuana Engelm.
Pino Prieto

TREE 15–25 m (50–80 ft) tall. LEAVES 6–14 cm (2.5–5.5 in) long, 3 per fascicle. SHEATH deciduous. CONE 4–6 cm (1.5–2.5 in) long, ovoid, persistent.

Pinus chihuahuana, a tree of dry, rocky hillsides, bears cones that require three years to mature. Persistent, single or whorled, the ovoid cones remain attached to the tree by their short peduncles for such a long period that the tree appears always to have cones. Bark of a young tree is grayish-red; deep ash-colored fissures form wide plates in the thick, blackish bark of the mature tree. Pale green, rather stiff leaves are usually in fascicles of three, although groups of two, four, or five may be found. The reddish-brown sheath is deciduous.

This pine, an important source of turpentine, grows in the western mountains from Sonora and Chihuahua south through Durango and Zacatecas to Nayarit and Jalisco. It is closely related to *P. leiophylla* and has been recognized as a variety of that species by some authors. It differs in having stiffer needles and fascicles of three, instead of five, leaves.

Pinus leiophylla Schl. and Cham.
Pino Chino or Ocote Chino

Similar to *P. chihuahuana,* this species also appears to have cones at all times. The three years required for maturation means that ovoid cones in various stages of development are visible at any season. Gray-green leaves are in fascicles of five with deciduous sheaths. The needles are very fine and delicate, characters readily recognizable when *P. chihuahuana,* with its shorter, stouter needles, is available for comparison. The reddish bark is thin at first, becoming coarser at an early age and separating in deciduous scales.

Pinus leiophylla, also an important source of turpentine, is found farther south than is *P. chihuahuana.* Although populations grow in Chihuahua, it becomes more common in the southern states of Michoacán, Puebla, and Veracruz.

PINACEAE—PINE FAMILY

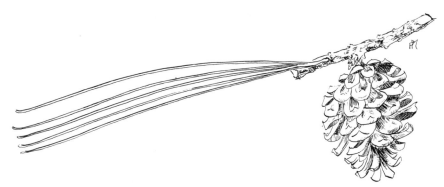

Pinus leiophylla
TREE 15–25 m (50–80 ft) tall. LEAVES 8–15 cm (3–6 in) long, 5 per fascicle. SHEATH deciduous. CONE 4–6 cm (1.5–2.5 in) long, ovoid.

Pinus lumholtzii Rob. & Fern.
Pino Triste

Long drooping needles, a natural growth form, present the appearance of a tree badly in need of water and have given rise to the common name of "pino triste" (sad pine). The long, clear green needles in fascicles of three at first are wrapped in long, papery, reddish-brown sheaths, which are promptly deciduous. Ovoid cones are on conspicuous peduncles, which fall with the deciduous cone. Common on young trees is a thin, scaly, chestnut or red brown bark, which becomes thick, dark, and rough as the pine ages.

A pine of the western mountains at elevations from 1600 to 3000 m (5200–9750 ft), *P. lumholtzii* is not of economic significance, for it produces very little low-grade turpentine. The hard wood is used for the manufacture of musical instruments. This pine occurs from Chihuahua south to Guanajuato and Jalisco, and has often been confused with the eastern weeping pine, *P. patula*, from which it differs in having the deciduous fascicle sheath, more strictly pendant leaves, and smaller cones.

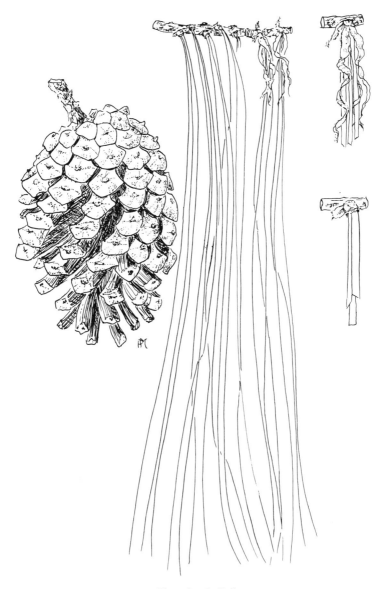

Pinus lumholtzii
TREE 15–20 m (50–65 ft) tall. LEAVES 20–30 cm (8–12 in) long, 3 per fascicle. SHEATH 30 mm long, deciduous. CONE 4–6 cm (1.5–2.5 in) long, ovoid; peduncle 1–1.5 cm (.5–.75 in) long.

PINACEAE — PINE FAMILY

Pinus montezumae Lamb
Pino Moctezuma

Pinus montezumae, a large tree with stiff needles on thick, heavy stems, has five dark green needles to a fascicle; each fascicle is surrounded at the base by a persistent sheath, which is brown at first but soon turns black. The branches supporting the leaves are of large diameter and are greatly roughened by the prominent scale leaves at the base of each fascicle. A thick, rough, cracked, reddish-brown bark is characteristic of this tree from youth to maturity. Ovoid or oblong-ovoid cones borne in pairs or in groups of three are sessile or on short peduncles. The cones fall at maturity, commonly leaving the peduncle on the branch.

An important source of lumber, this pine grows at elevations of 2500–2800 m (8200–9200 ft). It ranges as far north as Coahuila, Nuevo Leon, and Tamaulipas, and from there south to Chiapas and Guatemala.

Pinus oocarpa Schiede
Pino Prieto or Pino de Ocote

This highly variable pine is abundant, especially in southern Mexico. One distinction is the cone, which grows on a long persistent peduncle, either singly or in pairs. The cones are slow to mature and tend to remain on the tree for extended periods. Fascicles of five, long, slender, stiff or flexible, green needles are normal, but variety *trifoliata* has only three needles per fascicle, and var. *microphylla* has needles much shorter than normal. A prominent, persistent sheath enclosing the base of each fascicle develops horizontal striations as it matures.

Important as a timber tree in southern Mexico and Guatemala, where it may form almost pure stands, *P. oocarpa* ranges northward along both sides of the western Sierras to southern Sonora. Abundant turpentine is produced by this species, especially in dry weather.

Pinus montezumae
TREE to 30 m (100 ft) tall. LEAVES 15–35 cm (6–14 in) long, 5 per fascicle. ▶
SHEATH 15–25 mm long, persistent. CONE 8.5–15 cm (3.25–6 in) long, deciduous; peduncle 1–1.5 cm (.5–.75 in) long.

PINACEAE — PINE FAMILY

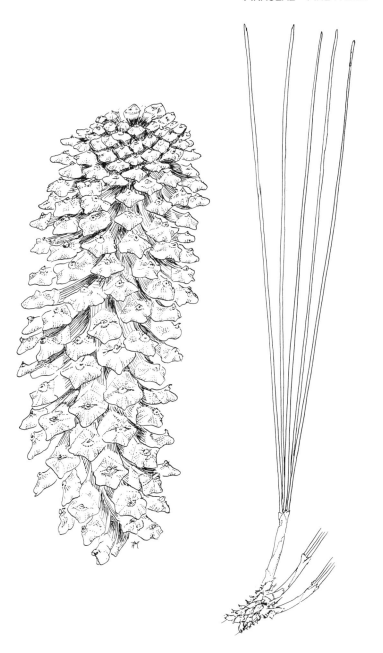

PINACEAE — PINE FAMILY

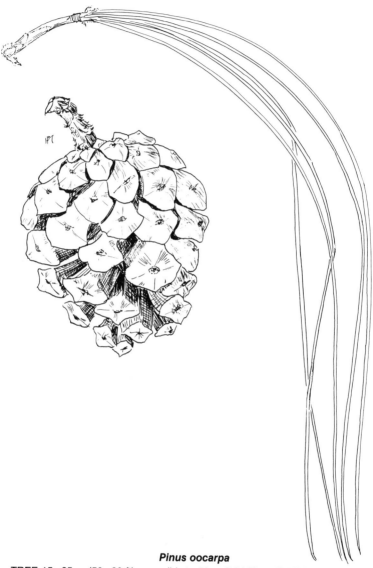

Pinus oocarpa

TREE 15–25 m (50–80 ft) — possibly to 50 m (164 ft) — tall. LEAVES 18–28 cm (7–11 in) long, 5 per fascicle. SHEATH 20–30 mm long, persistent. CONE 4–10 cm (1.5–4 in) long, ovoid or ovoid conic; peduncle 2–3 cm (.75–1.25 in) long.

Pinus patula Schl. and Cham.
Ocote Colorado

Recognizable by its slender, gracefully drooping leaves, *Pinus patula* is a tree of the warm humid areas of central and eastern Mexico. Fascicles of three clear, bright green needles are most common, although clusters of four or, rarely, as many as five may form. The drooping form of the delicate needles resembles that of *P. lumholtzii*, but the persistent grayish-brown sheath and strictly pendant position of the fascicles makes *P. patula* readily recognizable. The bark also aids in recognition: the upper trunk and branches are red and scaly even on young parts. Clusters form of three to six sessile or subsessile cones, which are ovate-conic, oblique, reflexed, and so tenaciously persistent that the cone may become embedded in the stem as the tree grows.

Pinus patula, often cultivated for its graceful appearance, is native in the states of Queretaro, Hidalgo, Veracruz, Puebla, and México. Its soft workable wood is used in making boxes. The color of the upper trunk prompts the common name "ocote colorado" (red pine).

Pinus teocote Schl. and Cham.
Pino Rosillo

Widely distributed and variable, *Pinus teocote* produces fascicles of three heavy, stiff, bright green or yellow-green needles subtended by a persistent, dark brown sheath. Small, ovoid or ovoid-conic, symmetrical cones, formed singly, in pairs, or in small clusters, soon fall, with their short peduncles attached. Young trees of this pine have a thin red bark, which at maturity becomes gray and rough with large longitudinal plates. *P. teocote* occurs from Chihuahua, Coahuila, Nuevo Leon, and Tamaulipas south along both the eastern and western mountains, at elevations of 1300–3400 m (4300–11200 ft), to Oaxaca and Chiapas. It produces good quality lumber and abundant turpentine.

PINACEAE — PINE FAMILY

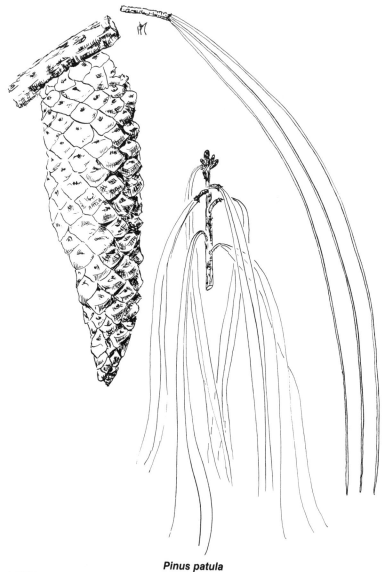

Pinus patula

TREE 10–25 m (32–80 ft) tall. LEAVES 15–30 cm (6–12 in) long, pendant, 3 per fascicle. SHEATH 10–15 mm long, persistent. CONE 6–11 cm (2.5–4.25 in) long, persistent, clustered.

PINACEAE—PINE FAMILY

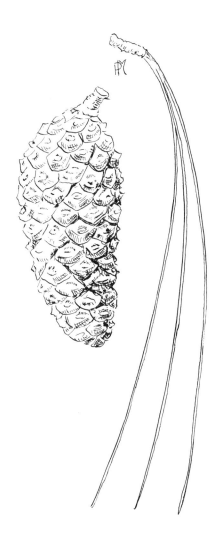

Pinus teocote

TREE 10–30 m (32–100 ft) tall. LEAVES 10–20 cm (4–8 in) long, 3 per fascicle. SHEATH persistent, 5–8 mm long. CONE 4–7 cm (1.5–3 in) long, symmetric; peduncle 5–8 mm long.

Piperaceae—Pepper Family

Piper aduncum L.

Piper aduncum, one of the widely distributed species of a very large genus, is conspicuous for its long, curved spike of very tiny flowers attached opposite to the nearly sessile leaves. Both the individual flowers and the berry fruits are so small as to be inconspicuous, but when massed into the spike inflorescence the flowers take on a pale green or greenish-white color. The leaves, pinnately veined with three to five lateral nerves on a side, are lance-oblong, with an abruptly attenuated apex and often an unequal base.

More commonly found in moist thickets, and particularly in areas of second growth, this species ranges from San Luis Potosí and Nayarit south to Chiapas. The fruits are used to flavor foods, for they have the properties of black pepper (which is derived from *P. nigrum*).

Piper auritum H.B.K.
Momo or Acoyo

A common succulent herb, *Piper auritum* has broadly ovate to oblong-ovate leaves with deeply cordate, unequal bases supported on heavy winged petioles that clasp the stem. Large, thin, soft leaf blades are pinnately veined, with a few lateral veins. Opposite the leaves stand the single, straight, simple spikes of numerous tiny, pale green flowers. The straight spike of flowers and the winged petiole on the leaves distinguish this *Piper* from *P. aduncum*.

Widely distributed in moist or wet thickets, and often as second growth in cut-over areas, *P. auritum* grows from San Luis Potosí to Yucatán and south throughout Oaxaca and Chiapas into Central America. The leaves and stems have a strong odor and are used to flavor meat dishes.

PIPERACEAE — PEPPER FAMILY

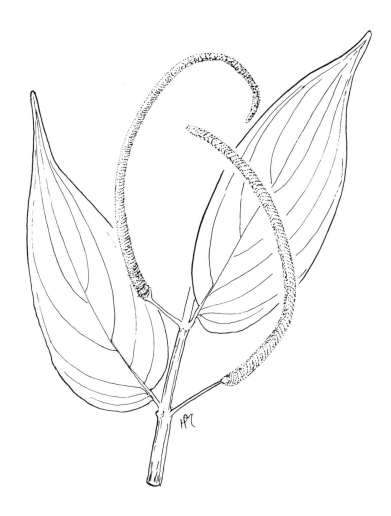

Piper aduncum

SHRUB or tree 1–5 m (3–16 ft) tall. LEAVES 13–20 cm (5–8 in) long, lance-oblong; pinnately nerved. INFLORESCENCE spike 10–13 cm (4–5 in) long, curved. FRUIT is a very small berry.

PIPERACEAE—PEPPER FAMILY

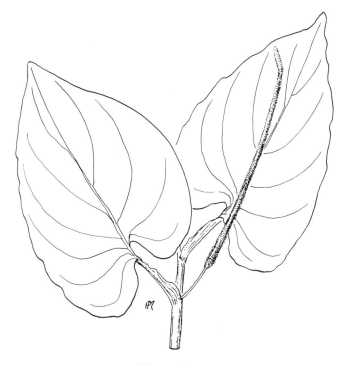

Piper auritum

HERB 2 m (6.5 ft) tall, succulent. LEAVES 60 cm (32 in) long, 35 cm (14 in) wide, broadly ovate; petiole winged. INFLORESCENCE 20–25 cm (8–10 in) long, simple spike opposite the leaf.

Polemoniaceae—Phlox Family

Cobaea scandens Cav.
Campanilla

Graceful, purple or green, bell-shaped flowers on long pedicels arising from the axils of the leaves make this vine a favorite plant for cultivation. A large campanulate corolla, its five frilly lobes slightly recurving at the tips, changes from green when young to violet upon

POLEMONIACEAE—PHLOX FAMILY

Cobaea scandens
VINE to 7.5 m (25 ft) long, climbing by leaf tendrils. LEAVES compound; leaflets 5—6 cm (2—2.5 in) long, ovate, acute. FLOWER 5—6 cm (2—2.5 in) long, bell-shaped, green, purple; pedicels 15—20 cm (6—8 in) long; 5 leaflike sepals. FRUIT 5 cm (2 in) long, capsule.

maturing. Resembling a clapper, five prominent stamens and stigma protrude beyond the lip of the bell. The five large, green, leaflike calyx lobes are strongly veined. Elliptic-ovoid fruits are large, glabrous, and full of thin, flat seeds.

This elegant vine climbs (*scandens* = climbing) by means of tendrils formed from modified apical leaflets. Alternate pinnately compound leaves have four to six leaflets and a terminal tendril.

Although restricted natively to Puebla and Veracruz, *Cobaea* has been widely cultivated and has become established in nonindigenous areas. In warmer regions it is treated as a perennial vine, but in colder localities it is used as a fast-growing annual, easily propagated by cuttings or seed. It flowers most of the year from February to November.

Loeselia mexicana (Lam.) Brand
Espinosilla

Loeselia, a stiff, colorful, shrubby perennial, has simple, lance-shaped, alternate leaves with sharp serrate margins. The numerous individual flowers have bright red corollas with five very short, recurving lobes. Commonly, five red stamens tipped with violet and the three-lobed stigma extend beyond the corolla, adding additional color to the small trumpetlike flower.

This handsome, winter-blooming shrub can be found from Sinaloa and Chihuahua south to Puebla and Oaxaca. A yellow-flowered variety grows on the Pedregal of Mexico City. *Loeselia* has been used for a variety of medicinal purposes; the crushed leaves served as a soap for early inhabitants.

Polygonaceae—Buckwheat Family

Antigonon leptopus Hook. & Arn.
Flor de San Diego or Queen's Wreath

Lacy masses of bright pink flowers direct attention to this clambering vine, which may reach the tops of the tallest trees and cover all other vegetation near it. Strong curling tendrils, formed at the ends of delicate

POLEMONIACEAE—PHLOX FAMILY

Loeselia mexicana
SHRUB to 1.5 m (5 ft) tall. LEAVES 2.5 cm (1 in) long, subsessile, glandular pubescent. FLOWER 2.5 cm (1 in) long, red, tubular.

POLYGONACEAE – BUCKWHEAT FAMILY

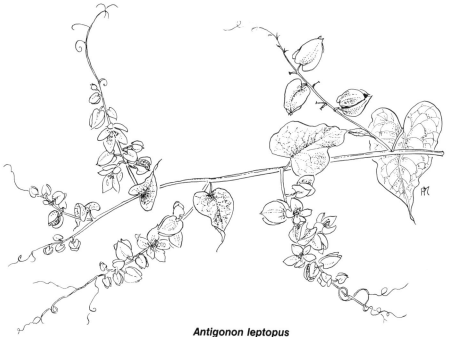

Antigonon leptopus
VINE to 10 m (32 ft) climbing by tendrils. LEAVES 4–20 cm (1.5–8 in) long, 2–12 cm (.75–5 in) wide, cordate. FLOWER pink; sepals 1 cm (.5 in) long. FRUIT achene 1 cm (.5 in) long.

flowering sprays, support this ornamental vine. Each flower has five heart-shaped, petal-like sepals, which at first are greenish but soon turn to pink, aging to magenta; there are no petals. After flowering, the sepals turn brownish and enlarge to cover the maturing fruit, which remains on the plant for some time. The fruit is shiny brown, glabrous, and three-angled. Alternately arranged leaves on slender petioles are heart-shaped and rough, with wavy margins and prominent veins. Edible tubers with a nutlike flavor are formed by the roots.

Widely cultivated and found throughout Mexico, this handsome vine has numerous common names. It is native from Chihuahua and Baja California south to Oaxaca. *Antigonon* flowers in summer, starting as early as April and continuing to November.

POLYGONACEAE – BUCKWHEAT FAMILY

Coccoloba uvifera (L.) Jacq.
Uva de la Playa, Uva de la Mar, or Sea Grape

"Uva de la mar" is a conspicuous seaside shrub or tree, by virtue of its large, roundish, thick leaves and elongated clusters of small white flowers or purple grapelike fruits. The plants vary in stature with the rigors of the habitat; they may be low, prostrate shrubs on windswept

Coccoloba uvifera
SHRUB or tree 9 m (30 ft) tall. LEAVES 20 cm (8 in) in diameter, thick, round. FLOWER inconspicuous, white. FRUIT 1 – 2 cm (.5 – .75 in) in diameter in grapelike clusters.

beaches, or spreading trees in a more favorable locale. Each large, leathery, round or kidney-shaped leaf has a membranous sheath encircling the stem. A main vein, often red or purple, contrasts with the very thick green leaf blade. Early Spaniards occasionally used sea grape leaves as writing paper; Standley, quoting the sixteenth-century writer Oviedo, notes that scratches on the leaf surface stand out in white, making writing easily legible.

Small, white, inconspicuous flowers borne in racemes are unisexual, with male and female flowers on different trees. They lack petals, but have five rounded calyx lobes. The male flower has eight stamens and a rudimentary pistil, whereas the female flower has a large pistil with three styles. Mature purple fruits are borne in long, dense clusters resembling a bunch of grapes. Each has a thin fleshy cover enclosing a large seed. It is edible, acidulous, and somewhat astringent; jelly and a winelike beverage can be made from the fruits.

Coccoloba uvifera (Latin *uva* = grape; *fera* = bearing) is found in coastal thickets from Tamaulipas and Sinaloa south through Central America. The wood is hard, reddish, and used for furniture; the bark of the tree yields a tannin.

Pontederiaceae—Pickerel-weed Family

Eichhornia crassipes (Mart.) Solms-Laub.
Violeta de Agua or Water Hyacinth

"A beautiful noxious weed," according to some botanists, *Eichhornia crassipes*, the "water hyacinth," can completely cover an aquatic area with its floating plants. Spreading, funnelform flowers borne loosely in terminal spikes, which rise conspicuously above the bulbous foliage, form a sea of light blue or violet. The six perianth parts are of almost equal size; only the upper lobe is slightly larger and highlighted in the center by a bright yellow spot surrounded by a dark blue patch. Three prominent, exserted stamens curve upward, while three others hide. The leathery leaf consists of an orbicular blade and an inflated petiole,

PONTEDERIACEAE — PICKEREL-WEED FAMILY

Eichhornia crassipes

HERB to 30 cm (12 in) tall, stemless; floating rootstocks. LEAVES to 15 cm (6 in) long, ovate to orbicular; petioles inflated at base. FLOWER to 5 cm (2 in) wide, 6 perianth parts, violet, in terminal spikes. FRUIT a capsule; many seeds.

which serves as a float for the plant. Abundant black roots hang from the water-level root crown.

The "water hyacinth" multiplies rapidly by vegetative means, spreading to clog waterways. *Eichhornia* can be found throughout Mexico in ponds, lakes, and slow-flowing streams.

Rhizophoraceae—Mangrove Family

Rhizophora mangle L.
Mangle Colorado or Red Mangrove

Rhizophora is a tree or large shrub with arching roots; it forms impenetrable thickets at tide level along muddy, brackish coastal waters. Its stiltlike roots enable this plant to survive in fluctuating tidal water where other trees would perish. Opposite, elliptic, blunt-pointed leaves on short petioles develop from long, tapering, apical buds. The slightly fleshy or leathery leaves are shiny green above, with a yellow-green underside. Small, delicately scented flowers in clusters of two to four on forked stalks are composed of four pale yellow sepals, four white petals, and a pistil with a two-lobed stigma. The persistent narrow sepals are leathery and recurve as the fruit develops. White petals, which soon turn brown, have white woolly hairs on the interior side.

Germination of the single seed while the fruit is still attached to the tree produces a conspicuous seedling, with a root that may reach 30 cm (12 in) in length as it hangs from the branch. Upon achieving sufficient weight, the germinated seed drops into the water or mud, where it becomes rooted to produce another tree.

In areas where the shoreline is building, the mangrove's stilted roots trap debris and hold the black mud in place, thereby extending the shoreline seaward. These roots also serve as a substrate for oysters and other aquatic animals. The reddish-brown, heavy wood is used for fuel and charcoal; leaves and bark are rich in tannin. Associated with red mangrove are *Avicennia*, *Laguncularia*, and *Conocarpus*.

Rhizophora extends from Sonora and Baja California south to Chiapas in a discontinuous, sporadic distribution. On the east coast it grows from southern Tamaulipas south to Yucatán and South America.

RHIZOPHORACEAE—MANGROVE FAMILY

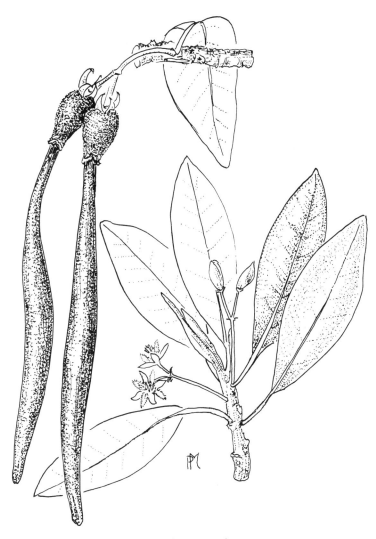

Rhizophora mangle
SHRUB or tree 3–8 m (10–33 ft) tall, gray bark. LEAVES 5–10 cm (2–4 in) long, opposite; terminal bud 2.5–5 cm (1–2 in) long. FLOWER 1 cm (.5 in) long, pale yellow, in pairs. FRUIT 1 seed, germinates on tree; root to 30 cm (12 in) long.

Rosaceae — Rose Family

Vauquelinia corymbosa Correa
Árbol Prieto

Water courses of the Chihuahuan Desert are home to this species. When its flat-topped clusters (*corymbosa*) of numerous, small, white, five-petaled flowers are in bloom it becomes a showy, densely foliated tree. Narrow evergreen leaves, several times longer than wide, have prominent marginal teeth. Small, woody, brown-fruiting capsules remain conspicuously on the tree long after they have opened to disperse seeds.

Standley reports that a yellow dye for goat skins is made from the wood and bark of *Vauquelinia*. The common name "guayule" has been applied to this species, but it is not to be confused with the more frequent application of that name to a completely different plant, *Parthenium argentatum*. *Vauquelinia* grows from Chihuahua, Coahuila, and Nuevo Leon south to Hidalgo.

Rubiaceae — Madder Family

Bouvardia ternifolia (Cav.) Schl.
Trompetilla

Bouvardia is a small shrub with two or three evergreen leaves at a node (*ternifolia* = leaves in threes). The petiole length (absent or short) and leaf shape (narrow to broadly lanceolate) vary from plant to plant and locality to locality. At the ends of the branches from March through November are cymes of red tubular flowers, the profusion of which make this shrub handsome, despite the smallness of the individual flowers. The corolla terminates in four short, spreading lobes; within the corolla tube are four stamens. A capsular, two-lobed fruit opens by valves, releasing the numerous, small, winged seeds.

"Trompetilla" (little trumpet) is a wide-ranging species common along roadsides and dry hills from Texas and Sonora south along the Sierra Madre Oriental and Sierra Madre Occidental to Veracruz and Oaxaca.

ROSACEAE—ROSE FAMILY

Vauquelinia corymbosa
TREE to 10 m (32 ft) tall. LEAVES 10–12 cm (4–5 in) long, narrow, evergreen. FLOWER small, 1 cm (.5 in) in diameter, white. FRUIT 5 mm long, brown, persistent, capsule.

RUBIACEAE—MADDER FAMILY

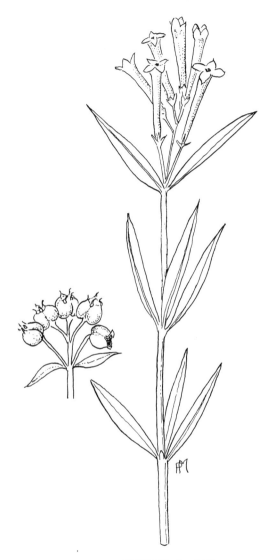

Bouvardia ternifolia

SHRUB or herbaceous plant to 1 m (3 ft) tall. LEAVES to 3 cm (1.25 in) long; commonly 3 at a node. FLOWER 2–3 cm (.75–1.25 in) long, red, tubular. FRUIT 5 mm in diameter, 2-lobed.

RUBIACEAE — MADDER FAMILY

Coffea arabica L.
Café

Coffea, an African plant whose fruit is in demand worldwide, is cultivated in the warm regions of Mexico and is also an established, naturalized escape. The attractive small tree has thin, short-petiolate, dark green, shiny leaves, opposite in arrangement, and lance-elliptic to oblong-oval in shape, with pointed tips. In the axil of each leaf are six buds, which can remain dormant or can develop into vegetative branches or flowers. Normally there are only four heavily scented white flowers in each inflorescence. The united sepals, with five tiny teeth, form a short cup at the top of the inferior ovary. A tubular corolla topped with five spreading lobes creates a star-shaped flower. Five stamens inserted on the corolla tubes have long anthers on slender filaments.

The flowering period is not constant, but all bushes burst into bloom the same day, as flower initiation is tied to length of day, temperature, and moisture. The self-pollinating blooms are short-lived, flowering in the morning and withering by evening. The inferior ovary remains on the plant to develop into a coffee berry, which takes nine months to mature and gradually changes in color from dull green, through yellow, to bright crimson. The fruit consists of a thin red skin covering a yellow pulp, in which two gray-green "coffee beans" are embedded. It is these seeds that are cured and roasted to produce coffee. Many processing factories in the coffee-growing area welcome visitors.

Coffea arabica is abundant in a wild state in parts of Veracruz and also occurs in Chiapas, Oaxaca and Tabasco. Finding the trees in bloom is a special delight, for the fragrance of the flowers is similar to that of the *Gardenia*, to which it is related.

Hamelia patens Jacq.
Coralillo or Chacloco

Hamelia, with its small red or scarlet flowers erectly blooming along one side of its multiple terminal inflorescences, can be found flowering somewhere in Mexico at any time of the year; not only is it widespread, but its blooming period is prolonged. The narrow, tubular, downy corolla ends in five short-pointed lobes, from which the five

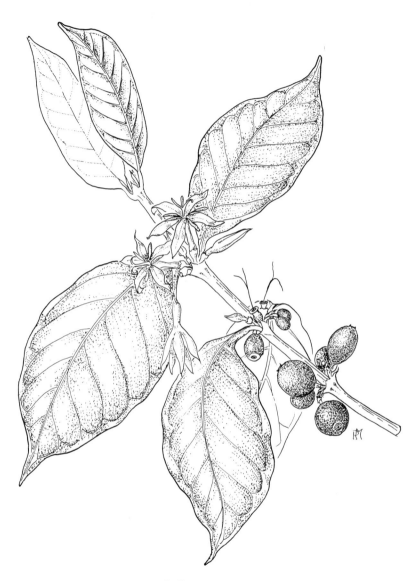

Coffea arabica

SHRUB or tree to 5 m (16 ft) tall. LEAVES 9–18 cm (3.5–7 in) long, 5 cm (2 in) broad, opposite. FLOWERS 4 per inflorescence; calyx minute; corolla tubular, 1.5–2 cm (.5–.75 in) long, lobes 1 cm (.5 in) long, white. FRUIT 10–16 mm long, globose or ovoid berry.

RUBIACEAE — MADDER FAMILY

Hamelia patens

SHRUB or tree 1–4.5 m (3–15 ft) tall, evergreen. LEAVES 5–15 cm (2–6 in) long, elliptic, opposite or in threes. FLOWER 2 cm (.75 in) long, tubular, scarlet. FRUIT a berry 6–10 mm in diameter.

yellow stamens barely show. At maturity the small, red or black fruit bears the ring of sepal lobes at the apex. This five-celled berry with many minute seeds is an edible but acidic fruit, from which a fermented drink can be concocted. Leaves, frequently growing three at a node, are elliptic, with prominent curving side veins. Dense hairs cover the veins of the lower surface of the leaf and are found along the petioles.

RUBIACEAE—MADDER FAMILY

The entire plant of *Hamelia*, including twigs, branches, flower clusters, flowers, fruits, petioles, and the midveins of leaves, all have a reddish or pinkish tinge.

Hamelia occurs along the east coast from Tamaulipas to Veracruz and Yucatán, into Oaxaca and Chiapas, continuing south through Central America. Although often planted as an ornamental, it sometimes is considered weedy because it grows in disturbed areas along the roadside, in waste places, and as an intruder in established hedges.

Randia echinocarpa Moc. & Sessé
Papache

If the white, tubular, star-shaped flowers or the peculiar medieval macelike fruits of *Randia echinocarpa* fail to attract the eye, its stout, rigid, gray branches, which terminate in four spines and a cluster of sessile leaves, will serve. The opposite, ovate or obovate leaves are densely hairy, at least along the veins on the lower surface. Fragrant white flowers with slender light green tubes are star-shaped as the five spreading lobes open to a flat-topped blossom. Five light brown stamens appear at the throat.

A green or yellow, pubescent, baseball-like fruit, armed with prominent tubercles, contains abundant seeds in an edible black pulp. The fruit commonly is offered in the market after the numerous, irregular, spiny tubercles are removed (*echino* = spine; *carpum* = fruit). The dark seeds and their enclosing black pulp form an inner ball, which will rattle when the fruit is mature.

This species is reported from dry hillsides of Sonora and Chihuahua south to Guerrero and Veracruz.

Sapindaceae—Soapberry Family

Dodonaea viscosa Jacq.
Gitarán or Hopbush

Most noticeable about this evergreen shrub are the papery, winged fruits that resemble the true "hop." The dark green foliage is extremely variable in form. Its sticky (*viscosa*) leaves tend to be very narrow in

RUBIACEAE—MADDER FAMILY

Randia echinocarpa

SHRUB or tree to 6 m (20 ft) tall; branch endings with 4 spines. LEAVES 3.5–8.5 cm (1.5–3.5 in) long, ovate or obovate. FLOWER 1–6 cm (.5–2.5 in) long, white, unisexual, fragrant. FRUIT 4.5–10 cm (2–4 in) in diameter; tubercles 1–3 cm (.5–1.25 in) long.

SAPINDACEAE — SOAPBERRY FAMILY

Dodonaea viscosa
SHRUB 1–3 m (3–10 ft) tall, evergreen. LEAVES 4–12 cm (1.5–5 in) long, sticky, variable, linear-oblanceolate. FLOWER 3 mm long, greenish-yellow. FRUIT 1.5–2.5 cm (.75–1 in) wide, winged.

the north, broader in the south. In all locations they are alternate, either sessile or with a short petiole. Small, inconspicuous, greenish-yellow flowers have three to five calyx lobes but no petals. The extremely large anthers of the five to eight stamens protrude from the calyx. Seed capsules, varying from green to red-brown to light tan upon maturity, have notches at both ends, two to three papery wings, and are present most of the year.

Dodonaea is found from Baja California to Nuevo Leon south throughout most of Mexico. Besides being a satisfactory ornamental, the plant has many uses: the leaves and bark for medicine, seeds for food, wood for tool handles and engraving, and fruits for flavoring beer.

Sapotaceae — Sapote Family

Bumelia lanuginosa (Michx.) Pers.
Coma or Chittamwood

Bumelia, a plant with deeply furrowed, dark grayish-brown bark, commonly grows as a shrub, but in favorable locations may become a tall tree. Elliptic to oblanceolate, leathery leaves are borne in fascicles or along sharply pointed spinelike branches. Among the leaf fascicles are clusters of small, fragrant flowers with five unequal sepals and five united petals. White, sterile stamens (staminodia) often are present. More conspicuous than the flowers are the black oval fruits, with an edible but unpalatable fleshy layer surrounding a single seed.

This tree grows most frequently along watercourses in upper desert regions of the northern tier of states from Sonora to Nuevo Leon. *Bumelia* belongs to the same family as "chicle" (*Manilkara achras*), the source of chewing gum, and produces a similar gummy exudate, which children of the area use as chewing gum.

Manilkara achras (Mill.) Fosberg
Chicozapote or Chicle

"Chicle," a very tall tree of the Tropical Evergreen Forest, has conspicuous, globose fruits and furrowed, dark brown bark, which may

SAPOTACEAE — SAPOTE FAMILY

Bumelia lanuginosa
SHRUB or tree to 20 m (65 ft) tall. LEAVES 2.5–8 cm (1–3 in) long, brown tomentose beneath. FLOWER small, white, sweet-scented. FRUIT 7–10 mm in diameter, fleshy, black.

be diagonally slashed for the collection of latex, its milky sap. The tree is handsome, with stiff, dark green, glossy, evergreen leaves clustered at the ends of stout branchlets. Inconspicuous, fragrant flowers growing singly in leaf axils have six greenish sepals and six white, united petals, plus six petal-like sterile stamens and a prominent pistil. The globose fruit's rough, thin, brown skin covers the yellow-brown, soft, sweet-at-

SAPOTACEAE — SAPOTE FAMILY

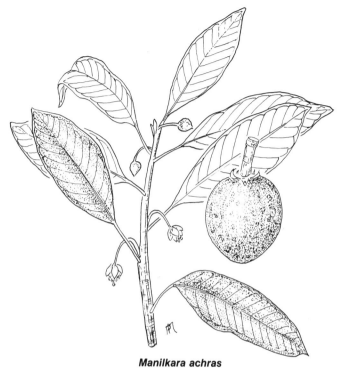

Manilkara achras
TREE to 40 m (130 ft) tall; bark dark brown. LEAVES 5–12 cm (2–5 in) long, elliptic-oblong. FLOWER 1 cm (.5 in) long, white. FRUIT 9 cm (3.5 in) in diameter, edible, ovoid.

maturity, edible flesh, and the numerous, small, variously flattened, shiny, black seeds. Immature fruits contain varying amounts of latex, which can be extracted by pressing.

Manilkara frequently is planted far from its native home in Chiapas, Tabasco, Campeche, Yucatán, and Quintana Roo. It has figured prominently in people's lives since early Mayan times; not only does it have large edible fruits, but it is the main source of the latex from which "chicle," the basic ingredient of chewing gum, is obtained. Both the Aztecs and Mayans were aware of the tree's latex properties, its extraction, and its uses. The wood is hard, heavy, and resistant to decay, and was used extensively by the Mayans in construction of their temples.

Scrophulariaceae — Figwort Family

Lamourouxia multifida H.B.K.

An herbaceous perennial, *Lamourouxia* produces racemes of red or orange flowers resembling certain species of *Penstemon*. The tubular corolla is bilabiate, with a large, unlobed upper lip that shelters the developing stamens, and a smaller, three-lobed, reflexed lower lip. Five united sepals produce a small cup topped with triangular, erect lobes. Delicate, deeply dissected, bipinnatifid leaves have two to five linear lobes on each side (*multifida* = many parted).

Growing in southern Sinaloa and Durango southward to Veracruz, Oaxaca, and Chiapas, this showy plant flowers most of the year, reaching its peak from August to December.

A second widely distributed species, *L. viscosa*, also with red flowers, has ovate to lanceolate leaves with serrate margins. Its veins usually show a prominent network on the lower leaf surface. This species grows in much the same area as does *L. multifida*, but extends northwestward into Sonora and Chihuahua and northeastward into Nuevo Leon and Tamaulipas. *L. viscosa* also flowers most of the year, with a peak season from September to December.

Leucophyllum frutescens (Berl.) I.M. Johnst.
Cenizo or Texas Ranger

Silver-gray shrubs that after a rain burst suddenly into lavender or purple bloom must be *Leucophyllum*. The ovate, sessile leaves grow compactly over the entire shrub and are densely covered with small star-shaped hairs, giving the entire plant a gray, felty appearance (*leuco* = white; *phyllum* = leaf; *frutescens* = shrubby). During the blooming period, flowers develop in great profusion from the axils of the leaves. Also densely hairy is the small, five-cleft calyx, which subtends the campanulate corolla. The five united petals of the corolla separate into recurving lobes, forming a bell-shaped flower that lasts only a day or two before dropping.

Natively, *Leucophyllum* is found from Texas south through Coa-

SCROPHULARIACEAE — FIGWORT FAMILY

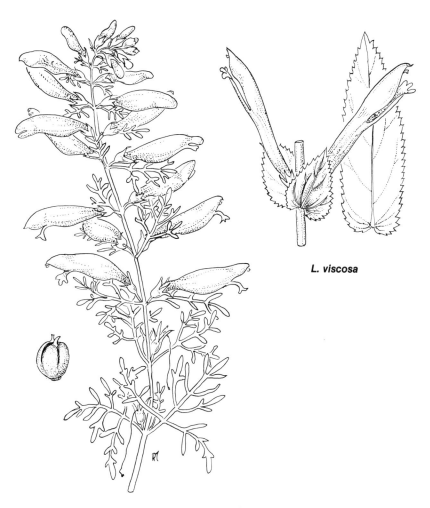

L. viscosa

Lamourouxia multifida
HERB to 2 m (6.5 ft) tall, perennial. LEAVES 20–55 mm long, 10–30 mm wide, dissected. FLOWER 22–55 mm long, red; calyx 6–11 mm long; inflorescence raceme. FRUIT capsule 9–11 mm long.

huila, Nuevo Leon, and Tamaulipas. This plant frequently is cultivated in the southwestern United States, where it is attractive as a rounded gray shrub, even before it responds to changing moisture by bursting into masses of purple flowers.

Leucophyllum frutescens
SHRUB to 2.5 m (8 ft) tall. LEAVES 2.5 cm (1 in) long, alternate, gray, felty. FLOWER 1.5−2.5 cm (.5−1 in) long, purple, tubular.

Maurandya erubescens (D. Don) Gray
Nen

Showy trumpet-shaped flowers, rose-red or purplish-rose (*erubescens* = blushing) on an herbaceous vine, may be seen on hillside rocks along highways. *Maurandya,* climbing by means of its twining pedicels and petioles, is a relative of the snapdragon. It has triangular-cordate leaves with coarsely dentate margins. Large irregular flowers arise singly on long graceful pedicels from the axils of the leaves. Five broadly ovate calyx lobes are erect during flowering, but spread outward as the fruit matures. Fused petals form an irregular, tubular corolla with an upper lip of two reflexed lobes and a lower lip of three somewhat erect lobes. Along the lower portion of the open throat are two plaits topped with yellow hairs. Included within the corolla are four fertile stamens in two lengths, along with a fifth, shorter, sterile one. The subglobose capsule disperses corky, tuberculate, slightly winged seeds through two irregular slits.

Maurandya is a graceful, colorful vine of the southern states; it was first collected near Jalapa, Veracruz, and is now reported also from San Luis Potosí, Morelos, and Chiapas.

Solanaceae — Nightshade Family

Cestrum nocturnum L.
Huele de Noche or Night Jessamine

Cestrum nocturnum, because of its pronounced fragrance, may be noticed before its small flowers are seen. Numerous white flowers in axillary or terminal racemes have narrow, tubular corollas tipped with five spreading, pointed, small lobes. Within the corolla tube are the five stamens and pistil, which after pollination produce white berries containing two cells but few seeds. The leaves of this small shrub are lanceolate to elliptic.

SOLANACEAE — NIGHTSHADE FAMILY

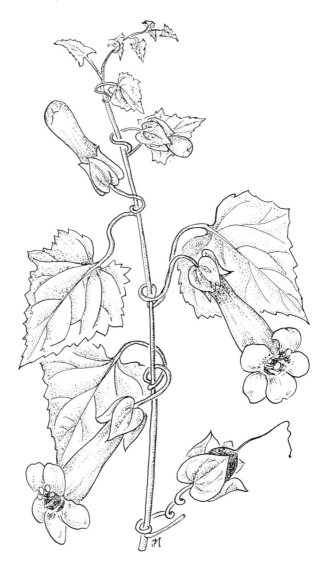

Maurandya erubescens
VINE herbaceous. LEAVES 15 cm (6 in) long, triangular-cordate. FLOWER 4–7 cm (1.5–3 in) long, rose-red, pedicels 2–6 cm (.75–2.5 in) long; calyx 1.5–2 cm (.5–.75 in) long, persistent. FRUIT 1.5 cm (.75 in) in diameter, subglobose.

SOLANACEAE — NIGHTSHADE FAMILY

Cestrum nocturnum

SHRUB 1–4 m (3–13 ft) tall. LEAVES 8–16 cm (3–6 in) long, 2.5–6 cm (1–2.5 in) wide, lanceolate to elliptic. FLOWER 20–25 mm long, greenish white, pale green; in axillary or terminal racemes. FRUIT 7–9 mm long, white berry.

Common natively throughout Mexico from Coahuila south to Veracruz, Guerrero, Oaxaca, and into Central America, this species of *Cestrum* often is cultivated for its sweet fragrance, which is especially noticeable at night. Many other species of *Cestrum* with larger and more showy flowers grow in Mexico, but none is as fragrant as *C. nocturnum*. The white berry is reported to be poisonous.

SOLANACEAE — NIGHTSHADE FAMILY

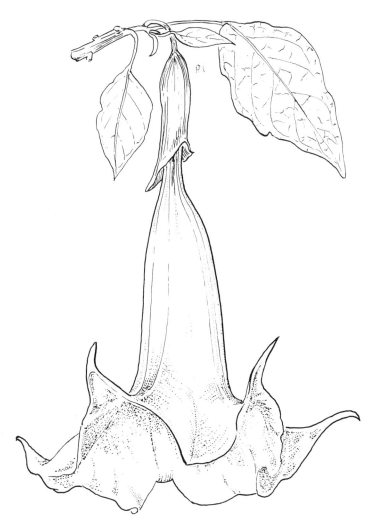

Datura candida

SHRUB or tree to 2.5 m (8 ft) tall. LEAVES 15–30 cm (6–12 in) long and wide, ovate to ovate-elliptic. FLOWER 27.5–37 cm (11–14 in) long, white; calyx 10–15 cm (4–6 in) long, spathelike; pedicel 3.5 cm (1.25 in) long. FRUIT 20 cm (8 in) long, 2 cm (.75 in) in diameter, unarmed.

SOLANACEAE — NIGHTSHADE FAMILY

Datura candida (Pers.) Safford
Floripondio

Large, white, pendant, trumpet-shaped flowers hanging like bells all over a shrub or small tree make it easy to recognize *D. candida* (*candida* = pure white). Five flaring lobes, each tipped with a long twisted tail, add distinction to this funnelform corolla. A tubular calyx, enwrapping the end of the corolla tube, is split along one side (spathaceous) and persists around the maturing, unarmed, spindle-shaped fruit. Ovate to ovate-elliptic, thick leaves have entire or wavy margins; most parts, particularly younger portions, are covered with small soft hairs, producing a velvety surface.

This showy plant grows from Sinaloa to Oaxaca and Veracruz. Although native to South America, it is popularly cultivated and can be seen in most frost-free areas. The fragrant, pure white flowers of *Datura* sometimes are placed beside the pillow of a person suffering from insomnia in the belief that the musky fragrance will induce sleep. All parts of *Datura* are poisonous.

Nicotiana glauca Graham
Gigante or Don Juan

Nicotiana is easily recognized by its tubular yellow flowers or by its long-petioled, ovate to oblong leaves, which are covered with a waxy bloom, giving them a blue-green color (*glauca* = with a bloom). This shrub or small tree often is scraggly in appearance, as its soft wood breaks easily. Numerous flowers at all seasons of the year bloom in loose, open, terminal panicles. The conspicuous yellow corolla forms a narrow tube extending to five small apical lobes. A tubular-campanulate calyx with five small teeth persists, partially covering the brown, ovoid seed capsule, which houses a multitude of minute, dark reddish-brown seeds.

Nicotiana, a native of South America, has successfully naturalized itself throughout Mexico from Baja California to Tamaulipas in the north, south to Oaxaca and Central America. This widely distributed,

SOLANACEAE—NIGHTSHADE FAMILY

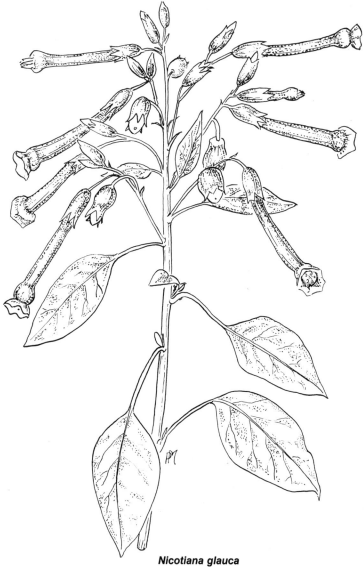

Nicotiana glauca

SHRUB or tree 8 m (25 ft) tall, evergreen. LEAVES 5–18 cm (2–7 in) long, glaucous, glabrous. FLOWER 3.5–4.5 cm (1.25–2 in) long, yellow, tubular. FRUIT 1 cm (.5 in) long, ovoid capsule.

spontaneously established, often weedy poisonous plant attracts hummingbirds, but is lethal to insects and repels livestock. It contains nicotine and other alkaloids and is closely related to tobacco (*N. tabacum*), as its name implies.

Solandra nitida Zucc.
Copa de Oro or Cup of Gold

"Copa de oro," with its glossy (*nitida* = shining) evergreen leaves and large flowers, is a striking, scrambling, woody shrub or vine. Handsome, solitary, chalicelike flowers appear most abundantly in the winter season from a loose, green, tubular, three- or four-lobed calyx. The fragrant showy corolla, narrowing to a funnelform tube, is golden yellow with five purple lines in the interior. The corolla lobes flare into

Solandra nitida

SHRUB or woody vine 18 m (60 ft) long. LEAVES 7–15 cm (3–6 in) long, elliptic; petiole 7 cm (3 in) long. FLOWER 18–25 cm (7–10 in) long, yellow; calyx 5–7 cm (2–3 in) long. FRUIT 4.5–5 cm (1.75–2 in) in diameter, globose.

SOLANACEAE – NIGHTSHADE FAMILY

a wide mouth, each lobe strongly reflexed. The opening of bud to blossom is so rapid that the movement of these magnificent, waxy flowers can be observed. Five prominent stamens curve upward from the throat. The large oval or globose fruit is edible.

A native of Puebla, Oaxaca, Veracruz, and Hidalgo, *Solandra* now is widely cultivated as an ornamental in Mexico and warmer parts of the United States.

SOLANUM SPECIES

Over 1700 species of herbs, shrubs, trees, and vines are included in the genus *Solanum*. Standley, in *Trees and Shrubs of Mexico*, lists 56 woody species for Mexico, and numerous herbaceous forms are found in the same area. Consequently, only a few of the species likely to attract attention are considered here.

The genus *Solanum* is not difficult to distinguish from other plants, for the five united petals form a more or less wheel-shaped corolla with definite lobes. The color may be white, yellow, blue-purple, or purple. Five stamens, which shed their pollen through apical pores, are somewhat united by their anthers and tend to stand prominently erect above the corolla. The fruit is a rounded berry with many seeds, and may be poisonous or edible. Some species are armed with prickles; others are unarmed, but may have abundant hairs. The following, divided into those with and without prickles, are a few of the species one might encounter in Mexico.

SOLANUM SPECIES WITH PRICKLES OR SPINES

Solanum hindsianum Benth.
Mariola

Solanum hindsianum is a much-branched shrub with scattered spines on its gray-green or yellow-green ovate leaves and stems. Blue to purple flowers have large wheel-shaped, but not deeply lobed, corollas with bright yellow anthers in the center. The calyx of five united sepals only half the size of the corolla is divided halfway to the base into narrow

SOLANACEAE — NIGHTSHADE FAMILY

Solanum hindsianum

SHRUB 1–3 m (3–10 ft) tall; spines 1.5 cm (.75 in) long. LEAVES 4.5 cm (2 in) long, felty. FLOWER 2.5–5 cm (1–2 in) in diameter, blue-purple. FRUIT a berry, 1–1.5 cm (.5–.75 in) in diameter, light green with dark green stripes.

linear lobes. A deeper colored midstripe extends from the center of the flower to the tip of each corolla lobe. Above the flower project the erect anthers, each about 1 cm long. Numerous seeds are contained in the glabrous, smooth, globose fruit, which is light green with darker green stripes.

This species is to be found on rocky hillsides of the desert regions of Sonora and Baja California. The common name "mariola," applied to this species in Baja California, is used for several other plants throughout Mexico.

SOLANACEAE — NIGHTSHADE FAMILY

Solanum hispidum Pers.
Sosa

Solanum hispidum, a shrub or small tree, is readily recognized by the stalked, red to reddish-brown, star-shaped hairs present on the stem and lower surface of the lobed, ovate leaves. Scattered among these hairs are stout, broad-based prickles. The white flowers in few-flowered inflorescences have deeply lobed corollas but still maintain the characteristic wheel-shaped bloom, with erect stamens projecting above the corolla. Small globose berries, black at maturity, represent the fruit of this species, which is reported from Michoacán to Veracruz and Chiapas.

Solanum madrense Fern.
Berenjena

Solanum madrense is a shrub densely covered with yellowish-brown, stellate hairs and armed with stout, straight or slightly curved, yellowish or pale brown prickles. The dark olive-green ovate leaves

Solanum hispidum
SHRUB or tree 1−5 m (3−16 ft) tall; spines small. LEAVES 30 cm (12 in) long, 20 cm (8 in) wide; star-shaped spines. FLOWER 20−28 mm in diameter, white. FRUIT a berry, 10−14 mm in diameter, black.

SOLANACEAE—NIGHTSHADE FAMILY

Solanum madrense
SHRUB 1–4.5 m (3–15 ft) tall; short, stout prickles 2–6 mm. LEAVES 5–8 cm (2–3 in) long, 4–13 cm (1.5–5 in) wide; prickles. FLOWER 2.5–3 cm (1–1.5 in) in diameter, white. FRUIT a berry, 8–15 mm in diameter, black.

show various degrees of lobing and may also be armed with prickles along the veins on the undersided. Inflorescences have few to many flowers with white, deeply lobed corollas. At maturity, the small globose fruits are black.

Flowering after both the spring and summer rains, this species can be found from southern Sonora and Chihuahua to Oaxaca, where its characters seem to merge with those of *S. torvum* of Veracruz and Chiapas; the latter extends its range into South America.

SOLANACEAE—NIGHTSHADE FAMILY

Solanum refractum Hook. & Arn.
Toronja

The large, globose, red or yellow fruits on this scandent shrub resemble tomatoes. *Solanum refractum* has short spines on the branches and broad-based prickles on the petioles and midribs of the large, oblanceolate or obovate, lobed or entire leaves. The flowers, in loose, many-flowered inflorescences, have white, deeply lobed corollas. Birds delight in the many-seeded edible fruit. This species of *Solanum* grows in the west central portion of Mexico from Sinaloa to Jalisco and Morelos.

Solanum refractum

SHRUB usually scandent, prickles. LEAVES 30 cm (12 in) long, oblanceolate. FLOWER 2.5 cm (1 in) in diameter, white. FRUIT a berry, 5 cm (2 in) in diameter, red.

SOLANACEAE — NIGHTSHADE FAMILY

Solanum rostratum Dunal
Duraznillo

Solanum rostratum, an annual herb, is densely covered with stellate hairs and well armed with numerous yellow spines on the stems, leaf petioles, midribs, and calyces. Leaves, oval to ovate in overall outline, are deeply lobed to pinnatifid with rounded segments. Showy yellow flowers in few-flowered racemes appear from May to September, to be followed by globose, brownish berries tightly enclosed by the spiny calyx, which doubles its size from flowering to maturation of the fruit. Within the berry are numerous black seeds.

S. rostratum is abundant as far south as the Valley of México, and is common as a weedy pest on overgrazed range land.

Solanum tequilense Gray
Huevo de Gato

Growing as a large herb or shrubby plant, *Solanum tequilense* is made conspicuous by its densely prickly stems. Broad-based spines also occur along the petioles and veins on both the upper and lower surfaces of the large, lobed leaves. Among the spines on all surfaces of the plant are abundant star-shaped hairs, some of which are stalked. White flowers in umbel-like inflorescences have deeply lobed corollas. Orange-red to red fruits densely covered with hairs help to distinguish this plant from many other species of *Solanum*. Named for the city of Tequila, Jalisco, where it was first found, this species now grows from Nayarit to Chiapas.

SOLANUM SPECIES WITHOUT PRICKLES OR SPINES

Solanum cervantesii Lag.
Hierba del Perro

Solanum cervantesii is a shrub or small tree with large, flat-topped clusters of white flowers. The spreading, lobed corollas are cut almost to the center. Elliptic-oblong or lanceolate leaves on long petioles are

SOLANACEAE — NIGHTSHADE FAMILY

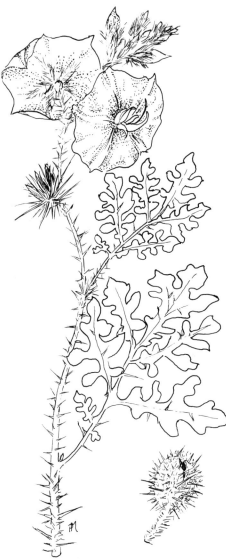

Solanum rostratum

HERB 40–80 cm (16–30 in) tall, annual; spines 5–15 mm long. LEAVES 12 cm (5 in) long, 5 cm (2 in) wide, 3 to 7-lobed. FLOWER 2–2.5 cm (.75–1 in) wide, calyx 5–7 mm long at time of flowering. FRUIT a berry, 9–10 mm in diameter, globose.

SOLANACEAE — NIGHTSHADE FAMILY

Solanum tequilense

HERB or shrub 1–1.5 m (3–5 in) tall, densely armed; spines 3 cm (1.25 in) long. LEAVES 15–45 cm (6–18 in) long, 4–15 cm (1.5–6 in) wide, lobed. FLOWER 3 cm (1.25 in) wide, white. FRUIT a berry, 2–2.5 cm (.75–1 in) in diameter, globose, orange-red to red.

SOLANACEAE — NIGHTSHADE FAMILY

glabrous on both surfaces, or the lower surface may have scattered hairs, which become dense along the veins. Children eat the comparatively small fruits, which are black at maturity, but care must be exercised because many species of *Solanum* are highly poisonous. This species grows from San Luis Potosí and Guanajuato south to Chiapas.

Solanum erianthum D. Don
Berenjena or Salvadora

The leaves, young stems, inflorescences, and fruits of *Solanum erianthum* are densely covered with stellate hairs. The abundance of hairs renders the ovate to ovate-elliptic leaves soft and velvety to the touch. Showy white flowers in large cymose inflorescences develop into globose, densely stellate, tomentose fruits that turn yellow at maturity.

Solanum cervantesii

SHRUB or tree 1−5 m (3−16 ft) tall, unarmed. LEAVES 7−20 cm (3−8 in) long, 2−4 cm (.75−1.5 in) wide; petioles 1−3 cm (.5−1.25 in) long. FLOWER 8−12 mm wide, white. FRUIT a berry, 5−10 mm in diameter.

SOLANACEAE — NIGHTSHADE FAMILY

Solanum erianthum

SHRUB or tree 2–8 m (6.5–25 ft) tall, unarmed. LEAVES 10–25 cm (4–10 in) long, 3–15 cm (1.25–6 in) wide; petioles 1–10 cm (.5–4 in) long. FLOWER 1–1.5 cm (.5–.75 in) wide, white, in cymes on stalks 3–12 cm (1.25–5 in) long. FRUIT a berry, 10–12 mm in diameter, yellow.

SOLANACEAE—NIGHTSHADE FAMILY

This common weedy species grows at elevations below 1000 m (3200 ft) throughout most of Mexico and into South America. Its range extends northward into southern Florida and southern Texas. *S. erianthum* has erroneously been called *S. verbascifolium* in much of the literature.

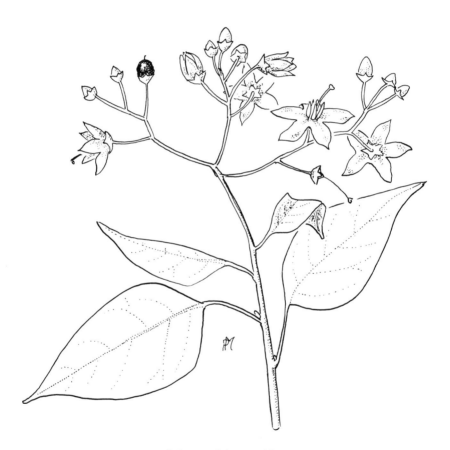

Solanum dulcamaroides

SHRUB viny, unarmed. LEAVES 5−14 cm (2−5 in) long; petioles 2−4.5 cm (.75−2 in) long. FLOWER 2−4 cm (.75−1.5 in) wide, violet; calyx 2−3 mm long. FRUIT to 1 cm (.5 in) in diameter, globose, red.

Solanum dulcamaroides Dunal

Solanum dulcamaroides, also called *S. macrantherum*, is a viny shrub with long-petiolate, broadly ovate, pubescent leaves. Violet, wheel-shaped flowers with deeply lobed corollas bloom in loose panicle clusters containing many blossoms. They are sweet-scented and showy, with conspicuous, broad, upright stamens. The fruit is a small, red, glabrous berry. *S. dulcamaroides* grows from Michoacán and Guanajuato to Veracruz and Chiapas.

Sterculiaceae — Sterculia Family

Chiranthodendron pentadactylon Larr.
Árbol de las Manitas or Macpalxochitl

One of the most intriguing flowers of Mexico resembles a small hand with long spreading fingers, a thumb, and a wrist. Appropriately its name in Spanish, Nahuatl, and Greek all allude to its unusual form: "arbol de las manitas" translates as tree of the little hand, and "macpalxochitl" as hand flower; in Greek, *chir* = hand, *anth* = flower, *dendron* = tree, *penta* = five, and *dactyl* = finger. The resemblance to a hand is so strong that no person can fail to be impressed by its form. Even more remarkable is the unique construction of the flower, for the prominent red "fingers" are not petals (a corolla is lacking), but a column of five fused stamens, which at the apex branches into five curved anthers, each ending in a slender, tapering, fingerlike tip. These bright red appendages extend prominently from the five-lobed, open calyx, which is brown tomentose on the exterior and brilliant dark red within. At the interior base of each thick, leathery (coriaceous) sepal is a prominent depression in which nectar accumulates. This honey is collected by natives. The entire large flower structure is subtended by two or three protective bracts. An oblong, woody, five-lobed fruiting capsule develops with small, shiny, black seeds.

Chiranthodendron, a large tree, has broad, ovate leaves on long petioles. Big leaf blades, palmately veined with five to seven shallow lobes, are so ample they often are used for covering and wrapping food. The dried or fresh flowers sold in herb markets are, by virtue of their

STERCULIACEAE — STERCULIA FAMILY

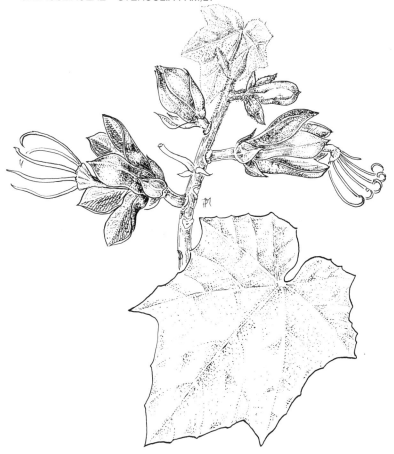

Chiranthodendron pentadactylon
TREE 12–30 m (40–100 ft) tall; 1–2 m (3–6.5 ft) in diameter. LEAVES 12–30 cm (5–12 in) long, 5–7 shallow lobes, palmately veined. FLOWER 3.5–5 cm (1.5–2 in) long; corolla lacking; stamens 5, red, fused into column, spreading. FRUIT 10–15 cm (4–6 in) long, woody capsule.

unique configuration, given religious significance, and are sought for their medicinal properties, which the Aztecs also knew.

The tree flourishes in the tropical rain forests of Chiapas, Oaxaca, Guerrero, Michoacán, and Morelos, where it grows natively, as well as in the National Palace garden and Chapultepec Park in Mexico City.

STERCULIACEAE—STERCULIA FAMILY

Guazuma ulmifolia Lam.
Guácima

Persistent warty black fruits make this small to medium-sized tree conspicuous. The alternate, ovate leaves, at times asymmetrical at the base, are finely serrated on the margins and resemble the leaf of an elm (*ulm* = elm; *folia* = leaf). Clusters of small, sweet-scented, creamy or yellow-greenish flowers arise in the angles at the bases of leaves. The hairy calyx is two- or three-parted, and the corolla has five short-clawed petals.

When immature the globose or oval fruit is green, mucilaginous, fleshy, and sweet. It is eaten raw or cooked and may be used as feed

Guazuma ulmifolia

SHRUB or tree 2–20 m (6.5–65 ft) tall. LEAVES ovate, serrated 6–12 cm (2.5–5 in) long. FLOWER small, cream colored. FRUIT 2–4 cm (.75–1.5 in) long, rough woody capsule.

for animals. Upon maturity (September to April) it turns into a distinctive, hard, black, warty capsule, which opens at the apex by pores to release the many brown seeds.

This is a tree of the stream banks and is abundant as second growth in clearings. It is found along the Pacific coast from southern Sonora to Chiapas, and on the Atlantic coast from southern Tamaulipas to the Yucatán Peninsula. *Guazuma* often is considered a weed tree, for it grows rapidly and quickly occupies cleared areas. The wood is little used, although furniture and miscellaneous small items are made from it. A strong fiber is obtained from the bark.

Theobroma cacao L.
Cacao

The sight of a small tree with overly large, ovoid-oblong fruit or miniature star-shaped flowers arising from the trunk or larger branches is an introduction to "cacao," the source of chocolate. *Theobroma cacao* was such a valuable discovery that it has been considered second only to gold in importance among all the treasures taken back to Europe by the early Spanish explorers.

"Cacao," an evergreen tree with a trunk rarely exceeding 15 cm (6 in) in diameter, has dark brown bark that is fissured and rough. Tiny star-shaped flowers on long pedicels are scattered in small inflorescences along the trunk, seemingly at random. The calyx consists of five narrow, pointed, widely spreading, pink lobes. Five yellowish petals make up the corolla; each petal is hood-shaped at the base, narrowed in the middle where it bends backwards, and spoon-shaped at the apex. The stamens lie hidden at the basal appendages of each petal. Standing erect in the center of the flower are five awl-shaped staminodia.

Hanging down from the tree trunk is the ovoid-oblong, fleshy yellow fruit with five conspicuous ribs. It contains many large purplish seeds, which the natives once used as a medium of exchange. A cacao bean yields chocolate and cocoa after being roasted and ground. Attractive dark green leaves hanging down on short petioles are alternate, elliptic-

STERCULIACEAE—STERCULIA FAMILY

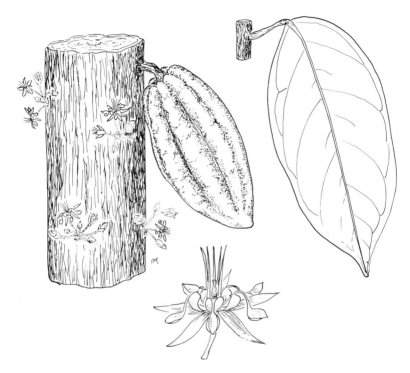

Theobroma cacao

TREE 6−8 m (20−25 ft) tall, evergreen. LEAVES 15−30 cm (6−12 in) long; petioles 2−3 cm (.75−1.25 in) long. FLOWER small along trunk; calyx lobes 6−7 mm long. FRUIT 30 cm (12 in) long, 10 cm (4 in) broad, ovoid-oblong capsule.

oblong or obovate-oblong with slightly thickened blades. The tree is sufficiently small to grow in the shade of other trees, where it is cultivated for its crop of seeds.

Although the native home of *Theobroma* is unknown, ''cacao'' has been cultivated since prehistoric times and is now distributed throughout the tropics. It can be seen in the forests and in cultivation in the warm humid areas of the states of Colima and Michoacán south to Chiapas, Tabasco, and Campeche.

Strelitziaceae — Strelitzia Family

Strelitzia reginae Banks
Ave del Paraíso or Bird of Paradise

Resembling a bird's head with a golden crest, *Strelitzia reginae* is a strikingly unusual plant in size, form, and color. Each large, boat-shaped, green bract is edged in pink and contains about six blossoms, which emerge one at a time from a tight slit in the bract to thrust three brilliant orange, pointed sepals upward. In vivid contrast is the central, bright blue, arrow-shaped portion of the flower, consisting of the modified petals, which stands erect between the intense orange sepals. Each bursting forth of this structural and colorful combination adds to the accumulating crest, until all the flowers contained inside the bract have emerged. Long, strong, petioled, dark green leaves, shaped like paddles, are a perfect background for this brilliant display of colors. Edges curved inward and with prominent veins, the leaves arise clumplike without a trunk.

A favorite in the gardens of tropical Mexico, *S. reginae* is a native of Africa. This beautiful flower was named in honor of George III's queen, Charlotte of Mecklenberg-Strelitz.

Taxodiaceae — Taxodium Family

Taxodium mucronatum Ten.
Sabino or Ciprés

Taxodium, a straight-trunked coniferous tree, grows along streams or in marshy areas. It has an enlarged base and brownish-red shredded bark. Needlelike yellow-green leaves form in a flattened plane; they are deciduous, and fall individually or as entire branchlets. Cones are of two types: small, ephemeral male cones on long slender spikes, and larger, globose, blue-gray female cones, which usually grow singly. The wood is soft and weak.

STRELITZIACEAE—STRELITZIA FAMILY

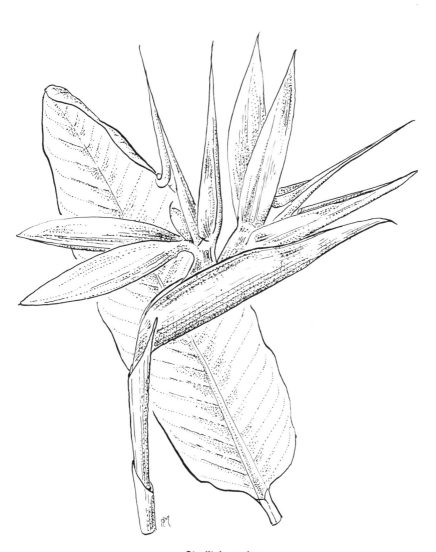

Strelitzia reginae

HERB to 1 m (3 ft) tall, perennial, trunkless. LEAF blades 25–45 cm (10–18 in) long, 10–15 cm (4–6 in) wide, paddle-shaped; petioles long. FLOWER 7.5 cm (3 in) long, orange and blue; bracts 15–20 cm (6–8 in) long. FRUIT a capsule.

TAXODIACEAE — TAXODIUM FAMILY

This species grows from Sonora to Coahuila south to Guatemala. A large specimen of *Taxodium* is the world famous "El Tule" in Santa Maria del Tule, south of Oaxaca. It is approximately 39 m (130 ft) tall, with a trunk 52 m (175 ft) in circumference. Other large specimens of this genus can be found in Chapultepec Park in Mexico City.

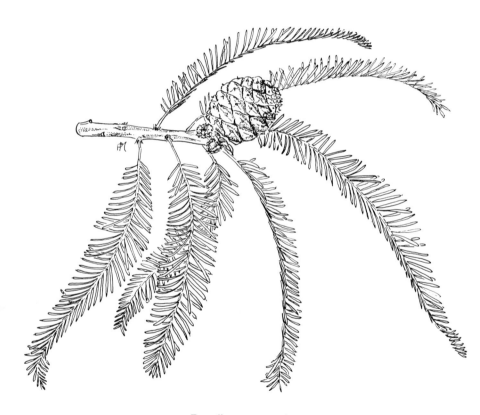

Taxodium mucronatum
TREE to 50 m (165 ft) tall; bark red-brown, shredded. LEAVES 6–12 mm long, needlelike, deciduous. MALE CONE 1.5–2.5 mm long, ovoid in spike. FEMALE CONE 1.5–2.5 cm (.5–1 in) long, subglobose. SEED 4–8 mm long, reddish-brown, angular.

Theophrastaceae — Theophrasta Family

Jacquinia pungens Gray
Pinicua

Jacquinia, a shrub or small tree, is likely to be felt before it is seen; its rigid, narrow, lanceolate leaves are tipped with a long, stiff, yellowish spine (*pungens* = sharp pointed). Glabrous evergreen leaves on many interlacing branches form a very dense crown. Small reddish-orange flowers, borne terminally in a corymblike inflorescence, are stiff and appear to be imitation flowers. The five-parted corolla spreads and recurves, each lobe edged with yellow. Within the corolla tube are five staminodia, which resemble the petals but are smaller and are inserted alternate to them. Five prominent stamens complete this stiff miniature

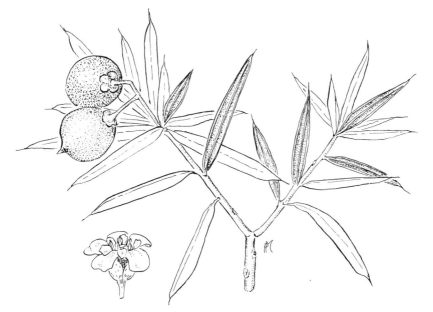

Jacquinia pungens

SHRUB or tree 1–4 m (1–13 ft) tall. LEAVES 3–6 cm (1.25–2.5 in) long with terminal spines. FLOWER 7–8 mm in diameter, stiff, red-orange. FRUIT globose, 1.5–2 cm (.5–.75 in) in diameter.

flower, which retains its form indefinitely when picked. The globose or ovoid fruit matures into a dry, woody ball with an elongated, sharply pointed tip. The fruits are used along the west coast for poisoning fish.

Jacquinia is found from Baja California and Sonora south to Chiapas, and on the east coast in Veracruz and the Yucatán.

Tiliaceae — Basswood Family

Heliocarpus attenuatus S. Wats.
Samo Babosa

Small, usually compressed, ovoid fruits with fringes of plumose hairs radiating like a halo around the sun introduce *Heliocarpus attenuatus*. The fruits develop from a dense terminal or axillary inflorescence of many small, inconspicuous flowers. The long, conspicuous, extended, attenuate apices on the ovate leaves resemble tails, thus accounting for the specific epithet *attenuatus*. The leaf margin is toothed, and five prominent veins expand into the blade from the top of the petiole. Stellate hairs sparsely or densely cover the leaf blades, veins, petioles, young growth, inflorescences, and fruits.

Heliocarpus attenuatus is a plant of northwestern Mexico, and is found in Sonora, Sinaloa, and Chihuahua. Flowering in August and September, its fruits can be seen as late as January.

The genus was named by Linnaeus (Latin *helio* = sun, *carpus* = fruit), who wrote, "Who could ever behold an almost round fruit bordered by a halo or rays without thinking of the sun as conceived by the painters?" The nine other species found in Mexico have the helioid fruits, varying in shape from globose to ellipsoid. Five species from the south central and southeastern portion of Mexico all bear the fruit on a short stipe (stalk), which also has plumose bristles. The five species from the northwest and southwest areas, including *H. attenuatus*, have sessile fruits. Although of little economic value, *Heliocarpus* bark yields a strong fiber used to make coarse cordage, ropes, mats, and baskets. The light, soft wood is suitable for floats and bottle stoppers.

TILIACEAE — BASSWOOD FAMILY

Heliocarpus attenuatus
SHRUB 2–3 m (6.5–10 ft) tall. LEAVES 3–10 cm (1.25–4 in) long, ovate attenuate. FLOWER 5–6 mm, inconspicuous. FRUIT 3 mm long, 2 mm wide, ovoid; 2 rows bristles, 6–7 mm long.

Ulmaceae—Elm Family

Celtis pallida Torr.
Granjeno Amarillo or Desert Hackberry

The "desert hackberry" is common in gravelly, well drained soils of desert areas throughout Mexico as far south as Oaxaca, where its intricate, dense, angular branching often forms impenetrable thickets. Small, ovate, evergreen leaves with coarsely toothed margins have three prominent veins. Paired, straight, thornlike branches arm the base of each leaf. Inconspicuous greenish-yellow flowers develop into edible, small, sweet, red, yellow, or orange fruits that are attractive to birds and other animals. Each glabrous, globose fruit has thin flesh and a large seed. *C. iguanaea*, a close relative found in much the same region, is distinguishable from *C. pallida* by its curved spines.

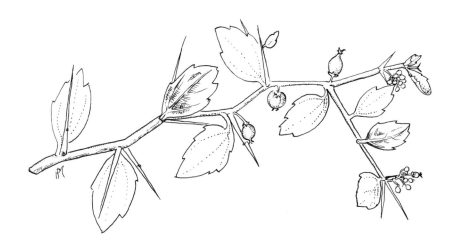

Celtis pallida

SHRUB 1–6 m (1–20 ft) tall, spiny, evergreen. LEAVES 1–3 cm (.5–1.25 in) long, 6–20 mm wide, oval. FLOWER inconspicuous. FRUIT 5–8 mm in diameter, edible.

Verbenaceae — Verbena Family

Avicennia germinans (L.) L.
Mangle Negro or Black Mangrove

This large shrub or tree is common in wet and swampy areas along both coasts. The young bark is smooth and gray, but with age it becomes dark brown, fissured, and scaly. Thick evergreen leaves on short petioles are opposite in arrangement. Clustered at the ends of the branches are small greenish to white flowers, which produce one-seeded, two-valved, capsular fruits that are somewhat compressed and oblique. As in the genus *Rhizophora*, the seed frequently germinates while the fruit is still attached to the tree. This tree also sends masses of aerial roots to the surface of the mud in which it grows.

"Mangle negro" (black mangrove) is found from Baja California and Tamaulipas south through Central America. The hard, dark brown wood is used for many purposes, and the bark is a source of tannin. The nectar of the flowers is sought by bees and is a source of a good grade of honey.

Avicenna is one of the four common shrubs or small trees occupying swampy tidal areas along both coasts; see also *Conocarpus*, *Laguncularia*, and *Rhizophora*.

Duranta repens L.
Duranta, Velo de Vivuda, or Espina Blanca

Duranta may occur as a shrub, a low tree, or even as a scrambling vine (*repens* = creeping). Its many small yellow berries persist for most of the year, and its slender branches often are weighted down by a profusion of fruit. Just as plentiful are the small blue, lilac, or white flowers that bloom in axillary or graceful terminal racemes. A five-lobed, tubular, salverform corolla and a small, green, tubular calyx make up the blossoms, which begin their extended flowering period in the spring. The tubular calyx with five small teeth turns yellow with age and continues to enlarge to enclose the yellow, globose fruit. A beak at the apex of the fruit develops from the small calyx lobes.

This plant may be unarmed or armed with straight spines. The

VERBENACEAE – VERBENA FAMILY

Avicennia germinans
SHRUB or tree 15 m (50 ft) tall. LEAVES 5–10 cm (2–4 in) long, elliptic, thick, evergreen. FLOWER 12 mm broad, greenish to white in terminal clusters. CAPSULE 2 cm (.75 in) long, 1.5 cm (.5 in) broad.

evergreen leaves, growing on short petioles, are ovate-elliptic in shape, and opposite or verticillate in position.

Frequently cultivated as an ornamental and often growing along roadsides, *Duranta* can be found from Baja California and Sinaloa south to Chiapas on the west coast, and from Tamaulipas to Veracruz, Puebla, and Yucatán on the east coast.

VERBENACEAE — VERBENA FAMILY

Duranta repens

SHRUB or tree to 6 m (20 ft) tall, armed or unarmed. LEAVES 1.5–5 cm (.5–2 in) long, opposite or verticillate. FLOWER 8–15 mm broad, salverform; in terminal or axillary racemes 6–15 cm (2.5–6 in) long. FRUIT 7–11 mm in diameter, round, beaked, yellowish.

VERBENACEAE — VERBENA FAMILY

Lantana camara L.
Lantana

Lantana is a small shrub, which at times may become trailing and grow through or over other vegetation. As it flowers most of the year, it is showy and colorful, with small red, orange, or yellow blooms forming dense axillary or terminal heads; these rise on stout peduncles above the foliage. Each tiny corolla is composed of a narrow tube with four spreading, irregular, flat-topped lobes. Within each flower head the marginal flowers open first and change colors as successive flowers bloom: yellow to orange, orange to red, red to reddish-purple. The ovate to oblong-ovate, deeply veined leaves with crenate or serrate margins have a pungent smell and are rough and hairy to the touch. Small, straight or curved prickles, either abundant or sparse, develop along the square stem between the pairs of opposite leaves. The blue or black, glabrous, small fruit is a juicy drupe with two seeds. These fruits may be poisonous.

Growing throughout most of Mexico, "lantana" frequently is cultivated as a garden shrub; it can become a serious weed in banana plantations and other agricultural areas.

Petrea volubilis Jacq.
Raspa-sombrero or Purple Wreath

Cascades of hyacinth-colored flowers hang in long, lacy racemes from a large woody vine called *Petrea*, another of the showy plants of Mexico that is commonly cultivated in warmer areas as an ornamental. Opposite, elliptic-oblong or obovate-oblong leaves on short petioles are leathery and rough. In the spring, blue flowers in graceful, pendant racemes arise from the leaf axils and commonly cluster near the ends of the stems. The conspicuous, colorful calyx has five narrow, blue, spreading, petal-like lobes that continue to enlarge as the fruit matures. It is this five-pointed, star-shaped calyx, which persists after the corolla has fallen, that makes this vine so attractive. Centered on the open calyx is the tubular, blue to purple corolla with five slightly irregular lobes, which resembles a tiny violet. It separates and falls from the calyx

VERBENACEAE — VERBENA FAMILY

Lantana camara

SHRUB 1–3 m (3–10 ft) tall, often scandent, armed. LEAVES 2–12 cm (.75–5 in) long, opposite, ovate to oblong-ovate. FLOWERS in dense heads; corolla tube 7–10 mm long; lobes 2–6 mm broad, yellow, orange, or red. FRUIT small, blue to black, juicy drupe.

VERBENACEAE — VERBENA FAMILY

Petrea volubilis
VINE 4–5 m (13–16 ft) long, woody. LEAVES 5–13 cm (2–5 in) long, opposite. FLOWER 1 cm (.5 in) blue-purple; racemes 8–20 cm (3–8 in) long. FRUIT 1.5–2 cm (.5–.75 in) long.

within a day or two of blooming. The inconspicuous fruit is completely enclosed by the lovely, blue, persistent, enlarging calyx.

Petrea flourishes in the warmer regions of Tamaulipas and San Luis Potosí south through Veracruz and Yucatán to Chiapas. It occurs natively in Oaxaca and Guerrero, but has spread to other areas by cultivation. In addition to its ornamental value, the tough stems (*volubilis* = twining) are used as ropes.

Zygophyllaceae — Caltrop Family

Guaiacum coulteri Gray
Guayacán

Guaiacum usually begins to bloom when leafless, but continues to flower after it is fully in leaf. The dainty, fragrant, bright blue flowers have five broadly ovate spreading petals, which narrow to a claw, exposing five sepals. Twice as many stamens as petals stand erect from the center of the open, slightly recurved corolla. The five-angled, green, fruit tinged with purple is conspicuous and persistent. Small wings develop on each angle. *Guaiacum* is crookedly branched, with closely crowded, pinnately compound, opposite leaves. The linear-oblong to elliptic oblong leaflets, even in number, are also in opposite arrangement. Arising from the axils of the leaves are the three- to twelve-flowered inflorescences. *G. coulteri* was named to honor the Irish botanist Thomas Coulter, who in the early 1800s was the first to botanize in Arizona.

Commonly found in Sonora, the range of *Guaiacum* extends south to Oaxaca. It is the resin from this plant that the Seri Indians of Sonora mix with clay to make a blue paint.

Larrea tridentata (DC.) Coville
Gobernadora, Hediondilla, or Creosote

Commonly referred to as "creosote bush" in the southwestern United States, this somewhat rounded evergreen shrub is one of the most common desert plants in the driest, hottest areas. The small, yellow-green, opposite, pinnately compound leaves have two sessile, basally-fused leaflets. A shiny coating of resin on the leaves imparts a

ZYGOPHYLLACEAE—CALTROP FAMILY

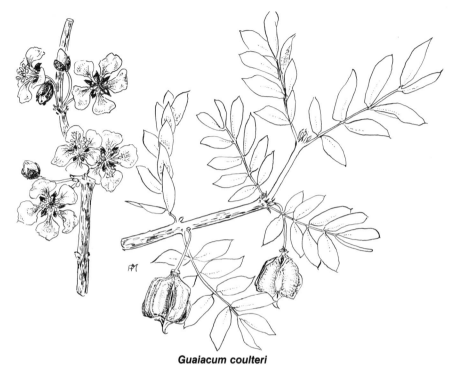

Guaiacum coulteri
TREE or shrub 2−8 m (6.5−25 ft) tall; bark light gray. LEAVES evenly compound; 6−8 leaflets 2−2.5 cm (.75−1 in) long. FLOWER 2−3 cm (.75−1.25 in) in diameter, blue. FRUIT 1.5 cm (.5 in) long 5 angled.

strong odor, resembling that of creosote, after a rain or when crushed. In the spring the bush is covered with an abundance of small, bright yellow, solitary flowers. Each dainty flower has five separate, somewhat twisted, spreading petals, with 10 stamens projecting erectly from the flat, open corolla. Globose fruits, which are densely covered with straight, radiating, white hairs, persist to give the bush the appearance of having silver balls for flowers.

Larrea covers vast expanses along the northern boundary from Baja California to Nuevo Leon, extending southward throughout the Sonoran and Chihuahuan deserts into the central area of Mexico, to Guanajuato, San Luis Potosí, and Hidalgo.

Larrea tridentata

SHRUB to 3.5 m (12 ft) tall, strong-scented. LEAVES opposite, pinnate, 2 leaflets 6–10 mm long. FLOWER 12–18 mm broad, solitary, 5 yellow petals not united. FRUIT 7–8 mm in diameter, globose, hairy.

Reference Material

Glossary

FLOWER STRUCTURE

The typical complete flower consists of four whorls of different parts attached to the enlarged end (receptacle) of a stemlike structure (pedicel). The outer whorl, commonly green and most leaflike, is the calyx, composed of individual sepals. The next whorl inward and alternating in position to the sepals is the usually showy, colored corolla, made up of petals within which in various numbers are the stamens (male reproductive parts). Each stamen may consist of a filament or stalk that supports the anther, the pollen producing structure. In the center of the flower is the female reproductive structure, composed of a stigma to receive the pollen, a style to support the stigma, and the ovary, which contains the ovule. Maturation of the ovule results in the formation of the seed; the maturing ovary becomes the fruit. It is the variation in size, color, shape, fusion, texture, ornamentation, and even the absence of these flower parts that provides the characteristics by which families, genera and species are recognized.

When the sepals, petals, and stamens are attached below the pistil, the ovary is considered superior; when these parts develop as though attached to the top of the ovary, the ovary is classified as inferior.

Achene — A small, dry, indehiscent, one-seeded fruit.
Acidulous — Slightly sour.
Acuminate — Gradually tapering to a point with somewhat concave sides; long-pointed.
Adventitious — Roots, buds, etc. out of the usual place.
Alternate — Placed singly at different heights on an axis or stem.
Anthesis — Time of opening of a flower; the time it is functional.

GLOSSARY

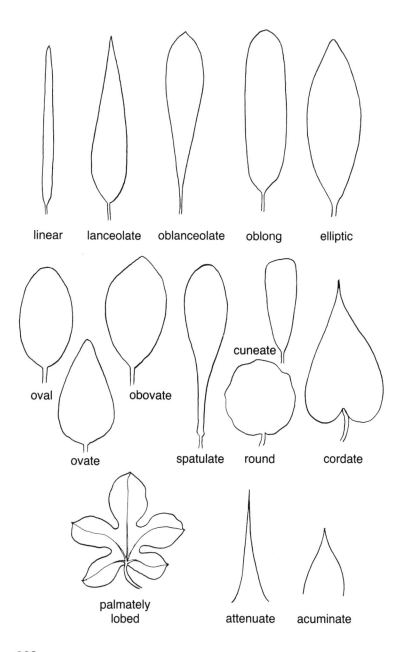

GLOSSARY

Anther — Pollen-bearing part of the stamen.
Apical — Situated at the top (apex).
Areole — In the Cactaceae, a small pit or raised spot, often bearing hairs, glochids, or spines.
Armed — With spines, thorns, prickles, barbs, etc.
Attenuate — Long, gradual taper, more gradual than acuminate.
Axil — Upper angle formed by a leaf or branch with a stem.
Axillary — Situated in the axil.
Axis — The main or central line of development (plural: axes).
Berry — A pulpy indehiscent fruit with one or more seeds but no true stone.
Bi- — Latin prefix meaning two, twice, or doubly.
Bifid — Two-cleft to about the middle.
Bilabiate — Two-lipped; often applies to calyx or corolla.
Bipinnate — When both primary and secondary divisions of leaf are divided (featherlike).
Bract — A modified leaf subtending the flower.
Bracteate — Provided with bracts.
Calyx — Outer whorl of a flower, composed of sepals.
Campanulate — Bell-shaped.
Capitate — Formed like a head; aggregated in dense cluster.
Capsule — Dry fruit that splits open at maturity.
Claw — Narrow constricted base of a petal.
Corm — A solid bulb-like part of a stem, usually underground.
Cordate — Heart-shaped.
Coriaceous — Leathery; tough.
Corolla — The inner whorl of a flower; the petals.
Corymb — A flat-topped or convex racemose flower cluster.
Crenate — Shallowly round-toothed, scalloped.
Cuneate — Wedge-shaped, triangular.
Cyathium — In Euphorbiaceae, cuplike structure bearing flowers.
Cyme — Determinate flower cluster with central flower opening first.
Deciduous — Falling off at end of growing period.

GLOSSARY

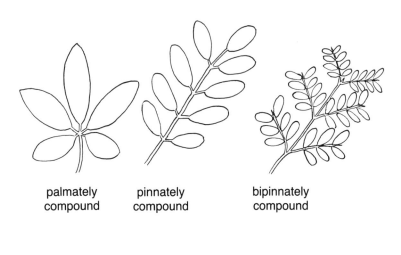

palmately compound   pinnately compound   bipinnately compound

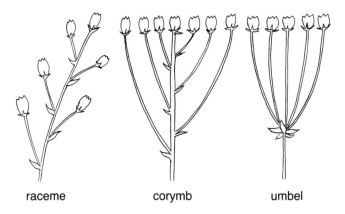

raceme   corymb   umbel

GLOSSARY

Dehiscent — Opening spontaneously when ripe to discharge contents; said of seed pod or stamen.
Dentate — With sharp, spreading teeth perpendicular to margin.
Denticulate — Minutely or finely dentate.
Dioecious — Female and male flowers on different plants.
Discoid — A round shape.
Discoid head — In Compositae, having only disk flowers.
Disk flowers — Tubular flowers in center of some Compositae heads.
Diurnal — Opening only during hours of daylight.
Drupe — An indehiscent, one-seeded stone fruit.
Entire — With a continuous margin; not in any way indented; whole.
Epiphyte — A plant growing on another plant but not parasitic.
Exserted — Protruding, as stamens projected beyond the corolla.
Fascicle — A small cluster of leaves, flowers, stems, or roots.
Filament — Thread, especially the stalk of the stamen supporting the anther.
Filiferous — With filaments or threads.
Follicle — Dry fruit that splits along one suture.
Funnelform — Gradually widening upward like a funnel.
Glabrous — Not hairy.
Glaucous — Covered with a bloom, a whitish material that rubs off.
Glochid — Minute spine or bristle in cactus areoles.
Helioid — Sunlike.
Herbaceous — Not woody; dying down each year.
Homology — Of similar structure.
Horny — Hard or dense in texture.
Inflorescence — General arrangement and disposition of flowers on an axis.
Indehiscent — Not splitting open, as a seed pod, anther or achene.
Inferior ovary — One that is below the attachment point of other flower parts.
Involucre — A whorl of bracts subtending a flower cluster, as in the heads of Compositae.

GLOSSARY

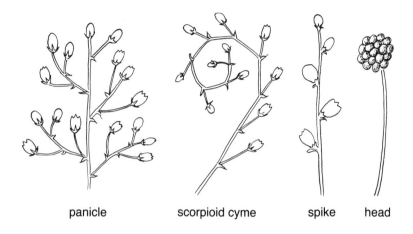

panicle   scorpioid cyme   spike   head

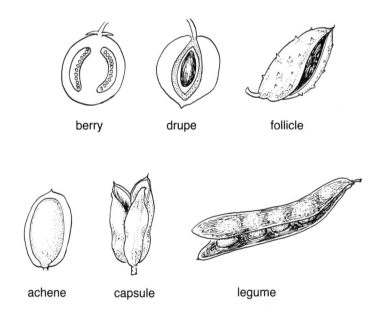

berry   drupe   follicle

achene   capsule   legume

GLOSSARY

Irregular — Showing a lack of uniformity; asymmetric.
Lanceolate — Several times longer than broad; widest below the middle.
Lenticel — Corky spot on young bark.
Limb — The expanded, free lobes of the fused petals.
Meristem — Undifferentiated tissue; the growing point.
Monoecious — With staminate and pistillate flowers on same plant.
Multicapitate — Many-headed.
Palmate — Hand-shaped; in a leaf, having lobes or divisions radiating from a common point.
Panicle — A multibranched inflorescence.
Paniculate — Borne in a pancile.
Parasite — An organism subsisting on another.
Pedicel — Stalk of one single flower in a flower cluster.
Peduncle — Stalk of a flower cluster.
Perfect — A flower having both stamens and pistil.
Perianth — Collectively, the sepals and petals of a flower.
Petiole — A leaf stalk.
Petiolule — Stalk of a leaflet.
Phyllary — An individual bract of the whorl of bracts subtending a flower cluster in the Compositae.
Pilose — Bearing soft, straight, spreading hairs.
Pinna — A leaflet or primary division of a pinnate leaf.
Pinnate — A compound leaf, having leaflets on either side of a common axis; featherlike.
Pinnatifid — Cleft into narrow lobes not reaching the midrib.
Pseudobulb — Swollen aerial stem of some orchids.
Puberulent — Minutely pubescent.
Pubescent — Covered with short, soft hairs.
Raceme — Simple elongated inflorescence with pedicelled or stalked flowers.
Rachis — Axis of a spike, of a raceme, or of a compound leaf.
Ray flower — In Compositae, marginal flowers as opposed to disk flowers; what one normally thinks of as a ''petal.''

GLOSSARY

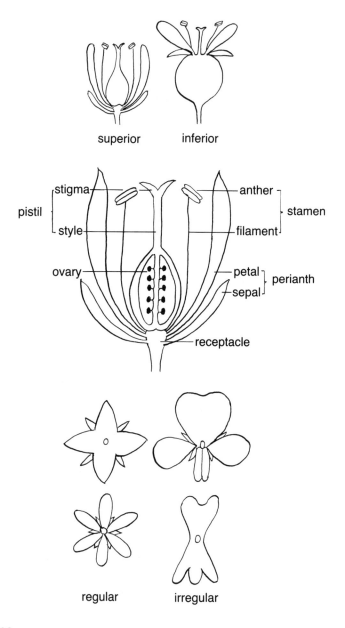

GLOSSARY

Receptacle — Structure to which flower parts are attached.
Regular — Radial symmetry, as in flowers, each series being alike.
Rhizome — Underground stem or rootstock with nodes, buds, scales.
Salverform — A corolla with slender tube expanding into flat limb.
Scabrous — Rough or gritty to the touch.
Scale — A vestigial leaf or bract, usually dry.
Scandent — Climbing in any manner.
Scorpioid — Like a scorpion tail.
Serrate — Saw-toothed, the sharp teeth pointing forward.
Sessile — Not stalked; attached directly by the base.
Sheath — Bracts or scales surrounding bundles of pine needles.
Sigmoid — Doubly curved like the letter *S*.
Spadix — A thick fleshy spike of flowers enveloped in a spathe.
Spathe — A broad sheathing bract enclosing or subtending a flower cluster or spadix.
Spatulate — Spatulalike; rounded above, narrowing to base.
Spicate — Having the form of or arranged in a spike.
Spike — An elongated inflorescence of sessile flowers.
Staminodium — A sterile stamen.
Stellate — Star-shaped.
Style — More or less elongated portion between stigma and ovary.
Suborbiculate — Almost circular in outline.
Subulate — Awl-shaped.
Superior ovary — Ovary above attachment point of other flower parts.
Terrestrial — Growing in the ground and supported by soil.
Tomentose — Covered with dense, matted, wool-like hairs.
Trifid — Three-cleft to about the middle.
Tubular — Shaped like a hollow cylinder.
Umbel — Flat-topped inflorescence; pedicels arising from a common point.
Verticillate — Arranged in whorls; a circular arrangement of similar parts at a node or axis.
Villous — Bearing long, soft, not matted hairs.

Selected References

Benson, Lyman and Robert A. Darrow. *Trees and Shrubs of the Southwestern Deserts*. 3rd ed. Tucson: University of Arizona Press. 1981.

Bravo H., Helia. *Las Cactaceas de México*. 2nd ed. con colaboración de Hernando Sanchez-Mejorado R. México: Universidad Nacional Autónoma de México. 1978.

Correll, Donovan and Marshall C. Johnston. *Manual of the Vascular Plants of Texas*. Renner: Texas Research Foundation. 1970.

Coyle, Jeanette and Norman C. Roberts. *A Field Guide to the Common and Interesting Plants of Baja California*. La Jolla: Natural History Publishing Co. 1975.

Gentry, Howard Scott. *Agaves of Continental North America*. Tucson: University of Arizona Press. 1982.

Gomez-Pompa, Arturo, ed. *Flora of Veracruz*. Vols. 1–41. Veracruz: Recursos Bióticos. 1978-1984.

Kearney, Thomas H. and Robert Peebles. *Arizona Flora*. 2nd ed. Berkeley: University of California Press. 1960.

Leopold, A. Starker. *Wildlife in Mexico*. Berkeley: University of California Press. 1972.

Little, Elbert, Jr. and Frank H. Wadsworth. *Common Trees of Puerto Rico and the Virgin Islands*. Washington: U.S. Department of Agriculture. Agric. Handbook 249. 1964.

————, Roy O. Woodbury, and Frank H. Wadsworth. *Trees of Puerto Rico and the Virgin Islands*. Vol. 2. Washington: U.S. Department of Agriculture. Agric. Handbook 449. 1974.

Martinez, Maximo. *Las Pináceas Méxicanos*. 3rd ed. México: Universidad Autónoma Nacional de Mexico. 1963.

SELECTED REFERENCES

Munz, Philip A. *A California Flora.* Berkeley: University of California Press. 1959.

O'Gorman, Helen. Edited by Ella Wallace Turok. *Mexican Flowering Trees and Shrubs.* México D.F.: Ammex Associados. 1961.

Pennington, T. D. and Jose Sarukhan. *Arboles Tropicales de México.* México D. F.: Benjamin Franklin. 1968.

Rzedowski, Jerzy. *Vegetacion de México.* México: Editorial Limusa, S. A. 1978.

Sanchez Sanchez, Oscar. *La Flora de Valle de Mèxico.* 6th ed. México D. F.: Editorial Herrero, S. A. 1980.

Shaw, George Russell. *The Pines of Mexico.* Boston: Ruiter & Co. 1909.

———, *The Genus Pinus.* Cambridge: Riverside Press. 1914.

Shreve, Forrest and Ira Wiggins. *Vegetation and Flora of the Sonoran Desert.* 2 vols. Stanford: Stanford University Press. 1964.

Standley, Paul Carpenter. "Trees and Shrubs of Mexico." *Contrib. U. S. Nat. Herb..* 23: 1-1721. 1920-26.

Wiggins, Ira L. *Flora of Baja Caifornia.* Stanford: Stanford University Press. 1980.

Williams, Louis O. "Orchidaceae of Mexico." *Ceiba* 2:1–321. Honduras. 1951.

Index

Abejón, 194
Abrojo, 113
Abutilon, 239
Acacia, 203–209; *cochliacantha*, 5, 206–207; *cornigera*, 206, 208–209; *farnesiana*, 203–204; *greggii*, 5, 203–206
Acalo Chuite, 280
Acanthaceae, 24–26
Acanthus Family, 24–26
Achimenes grandiflora, 175–176
Achiote, 91
Acoyo, 292
Acrocomia mexicana, 265–266
African Tulip Tree, 82
Agave, 26–38; *americana*, 28–30; *angustifolia*, 30–31, 38; *fourcroydes*, 32, 36; *lechuguilla*, 6, 32–34; *mapisaga*, 34, 36; *salmiana*, 34–36; *sisalana*, 36–37; *tequilana*, 30, 38–39
Agave Family, 26–46
Agavaceae, 26–46
Ageratum corymbosum, 132–135
Aguamiel, 28
Ahuejote, 72
Algarroba, 217
Amapa, 85
Amapa Prieta, 85
Amapola, 93
Amaryllidaceae, 46–51
Amaryllis Family, 46–51
Amate, 250, 253
Anacardiaceae, 51–56
Andira inermis, 217–219
Antigonon leptopus, 296–298

Apacahuite, 99
Apocynaceae, 56–63
Araceae, 64–65
Araliaceae, 66–67
Aretitos, 227
Arbol de Cuerno, 208
Arbol de las Manitas, 337
Arbol del Matrimonia, 118
Arbol Prieto, 304
Arbutus xalapensis, 6, 151–152
Argemone, 275–277; *mexicana*, 275–277; *ochroleuca*, 276
Aristolochiaceae, 66–69
Aristolochia grandiflora, 66–69
Arnotto, 91
Arnotto Family, 91–92
Arum Family, 64–65
Asclepiadaceae, 69–72
Asclepias curassavica, 69–70
Astianthus viminalis, 72–73, 74
Autumn sage, 185
Ave del Paraíso, 342
Avicennia germinans, 5, 131, 132, 302, 349–350
Ayal, 74

Bakeridesia notolophium, 239–240
Banana Family, 256–257
Barbas de chivato, 211
Barrel cactus, 106
Basswood Family, 346–347
Bayoneta, 262
Beard of the Goat, 211
Beaucarnea gracilis, 38–41
Bejuco Blanco, 145

373

INDEX

Bellotas, 167
Berenjena, 328, 334
Berrarco, 59
Bignonia Family, 72-91
Bignoniaceae, 72-91
Bird of Paradise, 192, 342
Bisnaga, 106
Bixa orellana, 91-92
Bixaceae, 91-92
Black Mangrove, 349
Boat-spined Acacia, 206
Bocconia arborea, 276-278
Bomarea hirtella, 46-48
Bombacaceae, 93-99
Bombax, 93-96; *ellipticum*, 93-94; *palmeri*, 6, 93-96
Bombax Family, 93-99
Bonete, 66
Boojum, 170
Boraginaceae, 99-101
Botoncillo, 130
Bougainvillea glabra, 145, 256-258
Bouvardia ternifolia, 304-306
Brahea dulce, 271-272
Brasil, 200
Bravo H., Helia, 104
Breadnut, 245, 248
Brittle Bush, 135
Brosimum alicastrum, 7, 245-248
Buckwheat Family, 296-300
Buddleia cordata, 227-230
Bugambilia, 256
Bull's Horn Acacia, 208
Bumelia lanuginosa, 313-314
Bursera, 101
Bursera fagaroides, 101-102
Bursera Family, 101-102
Burseraceae, 101-102
Buttonwood, 130-131
Byrsonima crassifolia, 234-235

Cabellos de Angel, 211
Cacahuananche, 221-222
Cacao, 340-341
Cactaceae, 103-126
Cactus Family, 103-126
Caesalpinia, 189-193; *cacalaco*, 189-190; *crista*, 191-192; *gilliesii*, 192; *pulcherrima*, 191-193

Caesalpinioideae, 189-203
Café, 307
Calabash, 74
California pepper tree, 56
Calliandra, 208-212; *anomala*, 211; *eriophylla*, 208-210; *houstoniana*, 211-212
Caltrop Family, 355-357
Campanilla, 294
Canastilla, 239
Candelabro, 123
Candelero, 140
Candelilla, 151-152
Candelillo, 157
Caoba, 245
Capulincillo, 242
Carica papaya, 126-127
Caricaceae, 126-127
Carnegiea gigantea, 104-106
Cardon, 116
Cascalote, 189
Cashew Family, 51-56
Cassia, 194-196; *biflora*, 194-195; *tomentosa*, 194-196
Castor bean, 160
Cat-claw Acacia, 203
Cecropia obtusifolia, 248-249
Cedrela odorata, 7, 242-245
Ceiba, 96-99; *aesculifolia*, 6, 96-97; *pentandra*, 96-99
Cedro, 242, 245
Celtis, 348; *iguanaea*, 348; *pallida*, 348
Cenizo, 316
Century Plant, 28
Cercidium, 197-198; *floridum*, 197; *microphyllum*, 197; *praecox*, 5, 197-198; *texanum*, 197
Ceriman, 64
Cestrum nocturnum, 319-321
Chacloco, 307
Chancarro, 248
Charresquillo, 208
Chicalote, 275
Chicle, 313, 315
Chicozapote, 313
Chilopsis linearis, 72, 74-75
Chimi, 128
Chiranthodendron pentadactylon, 337-338

INDEX

Chittamwood, 313
Cielitos, 132
Ciprés, 342
Cirio, 170
Clarín, 77
Clavellina, 93
Cobaea scandens, 294-296
Cocaiste, 271
Coccoloba uvifera, 299
Cochlospermaceae, 128-129
Cochlospermum vitifolium, 128-129
Coconut, 265, 266
Cocos nucifera, 265-267
Coffea arabica, 307-308
Cola de Mono, 149
Colorín, 221
Coma, 313
Combretaceae, 128-133
Combretum Family, 128-133
Combretum fruticosum, 128-130
Compositae, 132-144
Conocarpus erecta, 5, 130-132, 302, 349
Conostegia xalapensis, 242-243
Convolvulaceae, 145-149
Copa de Oro, 325
Copal, 101
Coralillo, 307
Corcho, 221
Cordia boissieri, 99-101
Corpus, 232
Coulter, Thomas, 355
Coursetia glandulosa, 220-221
Cousamo, 220
Coyol, 265
Coyol Real, 268
Creosote, 355
Crescentia, 74-77; *alata*, 74-77; *cujete*, 77
Cryptostegia grandiflora, 69-72
Cuajiote, 53
Cuajiote Amarillo, 101
Cuapinol, 200, 203
Cuerno, 69, 72
Cuerno de Toro, 208
Cuilumbuca, 217
Cunari, 170
Cup of Gold, 325
Cuphea jorullensis, 230-233

Cyatheaceae, 149-150
Cybistax, 82

Dama de Noche, 106
Datura candida, 322-323
Delonix regia, 191, 197-199
Desert Hackberry, 348
Desert Willow, 72, 74
Devil's Claw, 239
Distictis buccinatoria, 77-78
Dodonaea viscosa, 310-313
Dogbane Family, 56-63
Don Juan, 323
Duranta, 349
Duranta repens, 349-351
Duraznillo, 331

Echinocactus grandis, 106-107
Eichhornia crassipes, 300-302
El Tule, 344
Elm Family, 348
Encelia farinosa, 135-137
Encino Amarillo, 167
Encino Blanco, 162, 165
Encino Negro, 162
Encino Prieto, 165
Encino Rojo, 165
Enterolobium cyclocarpum, 211-213
Epidendrum parkinsonianum, 262-263
Epiphyllum oxypetalum, 106-109
Ericaceae, 151-152
Erythea, 275
Erythrina flabelliformis, 221-222
Espina Blanca, 349
Espinosilla, 296
Estrella, 48
Euphorbia, 151-161; *antisyphilitica*, 6, 151-153; *pulcherrima*, 154-155
Euphorbiaceae, 151-161
Evening Primrose Family, 260-262
Exogonium bracteatum, 145-146

Fagaceae, 162-169
Fairy Duster, 208
Fan Palm, 271-275
Feather Palm, 265-270
Ficus, 248-255; *carica*, 250; *cotinifolia*, 250-251; *glabrata*, 250-253;

INDEX

padifolia, 253–254; *petiolaris*, 253–255
Fig, 250
Figwort Family, 316–320
Flamboyán, 197
Flame Vine, 80
Flor de Arete, 260
Flor de Mayo, 48
Flor de San Diego, 296
Flor de San José, 194
Flor de Pascua, 154
Flor de Pato, 66, 69
Flor del Tigre, 178, 181
Flora de Madera, 230
Flora de Palo, 230
Floripondio, 323
Forget-me-not Family, 99–101
Four-o'clock Family, 256–260
Flourensia cernua, 6, 135–137
Fouquier, P. E., 175
Fouquieria, 170–175; *columnaris*, 170–171; *macdougalii*, 5, 171–173; *splendens*, 173–175
Fouquieriaceae, 170–175
Frangipani, 58
Frijolito, 225
Fuchsia fulgens, 260–262

Gallinita, 236
Galphimia glauca, 236–237
Gardenia, 307
Garumbullo, 109, 258
Gentry, Howard Scott, 28
Gesneria Family, 175–177
Gesneriaceae, 175–177
Gigante, 323
Ginseng Family, 66–67
Gitarán, 310
Gliricidia sepium, 221–223
Gloxinia, 175
Gobernadora, 355
Golden Heliconia, 256
Granadita, 46, 48
Granjeno Amarillo, 348
Gray Agave, 32
Guácima, 339
Guaiacum coulteri, 355–356
Guamuchil, 214
Guanacaste, 211

Guapinol, 203
Guayacán, 355
Guaycoyul, 268
Guayule, 304
Guazuma ulmifolia, 339–340
Guiro, 74

Haematoxylon brasiletto, 5, 200–201
Hamelia patens, 307–310
Heath Family, 151–152
Hediondilla, 355
Heliconia latispatha, 256–257
Heliocarpus attenuatus, 346–347
Hierba Azul, 24
Hierba del Cáncer, 230
Hierba Ceniza, 135
Hierba de la Raya, 147
Hierba de Leche, 69
Hierba del Perro, 331
Hopbush, 310
Hojasé, 135
Huajillo, 208
Huele de Noche, 319
Huevo de Gato, 331
Huisache, 203
Hura polyandra, 154–156
Hydrophyllaceae, 178–179
Hymenaea coubaril, 200–203

Idria columnaris, 170
Ingerto, 230
Ipomoea, 145–149; *arborescens*, 145–147; *pes-caprae*, 147–149
Iridaceae, 178–181
Iris Family, 178–181
Ironwood, 223
Izote, 44, 46

Jabillo, 154
Jacaranda, 78, 80
Jacaranda mimosifolia, 78–80
Jacobinia, 24
Jacquinia pungens, 345–346
Jatropha dioica, 6, 157–158
Jarilla, 142
Jícama Montes, 46, 48
Jumping Cholla, 112
Justicia spicigera, 24–25

INDEX

Kapok, 93, 96
Kohleria deppeana, 175–177
Labiatae, 181–189
Laguncularia racemosa, 5, 131, 132–133, 302, 349
Lamourouxia, 316–317; *multifida*, 316–317; *viscosa*, 316–317
Lantana, 352
Lantana camara, 352–353
Large-Thorned Acacia, 206–209
Larrea tridentata, 6, 355–357
Laurel Rosa, 56
Lechuguilla, 32, 34
Leucophyllum frutescens, 316–318
Leguminosae, 189–226
Liliaceae, 42
Lily Family, 42
Llamarada, 80
Llora Sangre, 276, 278
Lluvia de Oro, 236
Lobelia, 227
Lobelia Family, 227–228
Lobelia laxiflora, 227–228
Lobeliaceae, 227–228
Loeselia mexicana, 296–297
Logania Family, 227–230
Loganiaceae, 227–230
Loosestrife Family, 230–233
Loranthaceae, 230–231
Lysiloma divaricata, 5, 214–215
Lythraceae, 230–233

Macpalxochitl, 337
Madder Family, 304–311
Madre Cacao, 221–222
Madroño, 151
Magnolia Family, 232–234
Magnoliaceae, 232–234
Magnolia schiedeana, 232–234
Maguey, 28
Maguey de Pulque, 34
Maguey Mapisaga, 34
Maguey Tequilero, 38
Mahogany, 245
Mahogany Family, 242–246
Mallow Family, 239–240
Malpighia Family, 234–238
Malpighiaceae, 234–238
Malvaceae, 239–240

Mano de Leon, 66
Mangifera indica, 51–53
Mangle Blanco, 132
Mangle Colorado, 302
Mangle Negro, 349
Mango, 51
Mangrove Family, 302–303
Manilkara achras, 7, 313–315
Mariola, 326–327
Martynia Family, 239–241
Martyniaceae, 239–241
Mascagnia macroptera, 236–238
Mata Raton, 221–222
Matapalo, 253
Maurandya erubescens, 319–320
Melastomataceae, 242–243
Melastome Family, 242–243
Meliaceae, 242–246
Mescal, 28
Mescal Azul, 30, 38
Mescal Bacanora, 30
Mescal de Maguey, 30
Mesquite, 217
Mexican Bird-of-Paradise, 191–192
Mexican Poppy, 275
Milkweed Family, 69–72
Milla biflora, 48–49
Mimbre, 74
Mimosoideae, 189, 203–217
Mint Family, 181–189
Mirasol, 142
Mirto, 182
Mistletoe cactus, 118
Mistletoe Family, 230–231
Mohintli, 24
Mojagua, 239
Mojo, 245
Momo, 292
Monstera deliciosa, 64–65
Montanoa tomentosa, 138–139
Moraceae, 245–255
Morning-glory Family, 145–149
Mulberry Family, 245–255
Musaceae, 256–257
Myrtillocactus geometrizans, 109–110

Nacapuli, 250
Nanche, 234
Narciso Amarillo, 61

377

INDEX

Nen, 319
Nephelea mexicana, 149–150
Neobuxbaumia mezcalaensis, 111
Nerium, 56
Nerium oleander, 56–57
Nicotiana, 323–325; *glauca*, 323–325; *tabacum*, 325
Night Jessamine, 319
Nightshade Family, 319–337
Nopal de Castilla, 113
Nyctaginaceae, 256–260

Oak Family, 162–169
Oceloxóchitl, 178, 181
Ocote Chino, 283
Ocote Colorado, 289
Ocotillo, 170, 173, 175
Ocotillo Family, 170–175
Ojo de Venado, 191
Oleander, 56
Olneya tesota, 5, 223–225
Onagraceae, 260–262
Opuntia, 112–116; *ficus-indica*, 112, 113–114; *imbricata*, 113–116
Orbygna guacuyule, 268–269
Orchid Family, 262–263
Orchidaceae, 262–263
Oreopanax peltatus, 6, 66–67
Organ Pipe cactus, 123
Orejón, 211
Organo, 121
Ortega, 178

Pachycereus pectin-aboriginum, 116–117, 126
Palm Family, 264–275
Palma Blanca, 274
Palma China, 44, 46
Palma de Coco, 265
Palma Cristi, 160
Palma de Sombrero, 271
Palma Pita, 44
Palma Real, 271
Palma Redonda, 271
Palma Samandoca, 42
Palmae, 264–275
Palmito, 271
Palo Blanco, 145
Palo de Agua, 72

Palo de Cacique, 232
Palo de Rosa, 88
Palo del Diablo, 276
Palo Fierro, 223
Palo Mantescoso, 197
Palo Verde, 197
Papache, 310
Papaveraceae, 275–278
Papaya, 126
Papaya Family, 126–127
Papilionoideae, 189, 217–226
Parthenium argentatum, 6, 304
Pata de Paloma, 262
Pata de Vaca, 147, 149
Patito, 175
Pea Family, 189–226
Pedilanthus macrocarpus, 157–160
Peineta, 128
Pepper Family, 292–294
Pepper Tree, 53
Pereskia lycnidiflora, 118–119
Petrea volubilis, 352–355
Phaedranthus buccinatorius, 77
Phlox Family, 294–297
Pickerel-weed Family, 300–302
Pinaceae, 279–291
Pine Family, 279–291
Pinicua, 345
Pino Cahuite, 280
Pino Chino, 283
Pino de Ocote, 286
Pino Moctezuma, 286
Pino Prieto, 283, 286
Pino Rosillo, 289
Pino Triste, 284
Piñon, 280
Pinus, 6, 279–291; *ayacahuite*, 280–281; *cembroides*, 280–282; *chihuahuana*, 283; *leiophylla*, 283–284; *lumholtzii*, 284–285, 289; *montezumae* 286–287; *oocarpa*, 286–288 — var. *microphylla*, 286; var. *trifoliata*, 286 — *patula*, 284, 289–290; *teocote*, 289–291
Pipe Vine Family, 66–69
Piper, 292–294; *aduncum*, 292–293; *auritum*, 292–294; *nigrum*, 292
Piperaceae, 292–294
Pirul, 53

INDEX

Pisonia, 258–260; *aculeata*, 260; *capitata*, 258–260
Pitayo Dulce, 123
Pithecellobium dulce, 214–216
Platanillo, 256
Pluchea odorata, 138–140
Plumeria acutifolia, 6, 58–59
Pochote, 96
Poinsettia, 154
Polemoniaceae, 294–297
Polygonaceae, 296–300
Pontederiaceae, 300–302
Poppy Family, 275–278
Prickly Pear, 112, 113
Primavera, 82
Proboscidea fragrans, 239–241
Prosopis, 217–218; *glandulosa*, 217; *juliflora*, 5, 217–218; *velutina*, 5, 217
Pseudosmodingium pernicosum, 53–54
Psitticanthus calyculatus, 230–231
Pulque, 28, 34, 36
Purple Wreath, 352
Pyrostegia venusta, 80–81

Queen's Wreath, 296
Quercus, 6, 162–169; *albocincta*, 162–163; *candicans*, 162–165; *conspersa*, 165–166; *emoryi*, 165–167; *magnoliifolia*, 167–168; *pennivenia*, 169
Quiebracha, 214
Quisache Tepamo, 206

Rain-of-Gold, 236
Ramón, 245, 248
Randia echinocarpa, 310–311
Raspa-sombrero, 352
Red Mangrove, 302
Retama, 88, 194
Rhipsalis baccifera, 118–121
Rhizophora mangle, 5, 131, 132, 302–303, 349
Rhizophoraceae, 302–303
Ricinis communis, 160–161
Ricino, 160
Rosa Amarillo, 128
Rosaceae, 304–305
Rose Family, 304–305
Roseodendron donnell-smithii, 82–83

Royal Poinciana, 191, 197
Rubber Stem, 157
Rubiaceae, 304–311
Ruellia speciosa, 24–26

Sabal mexicana, 271–273
Sabino, 342
Sacqui, 32
Saguaro, 104, 106
Salix, 74
Salvadora, 334
Salvia, 181–189; *coccinea*, 182–183; *elegans*, 182–184; *gesneraeflora*, 185–186; *greggii*, 185, 187; *iodantha*, 185–189
Salvia Roja, 182, 185
Samo Babosa, 346
Samo Prieto, 220
San Pedro, 88
Sand Box Tree, 154, 156
Sangre de Drago, 157
Santa Maria, 138
Sapindaceae, 310–313
Sapotaceae, 313–315
Sapote Family, 313–315
Scheelea liebmannii, 268–270
Schinus molle, 53–56
Scrophulariaceae, 316–320
Sea Grape, 299
Senecio, 140–143; *praecox*, 140–142; *salignus*, 142–143
Shell Seed, 148
Shell Seed Family, 128–129
Siranda, 250
Sisal, 36
Soapberry Family, 310–313
Solanaceae, 319–337
Solandra nitida, 325–326
Solanum, 326–337; *cervantesii*, 331–334; *dulcamaroides*, 336–337; *erianthum*, 334–335; *hindsianum*, 326–327; *hispidum*, 328; *macrantherum*, 337; *madrense*, 328–329; *refractum*, 330; *rostratum*, 331–332; *tequilense*, 331–333; *torvum*, 329; *verbascifolium*, 336
Sophora secundiflora, 225–226
Sosa, 328
Sotolin, 38

INDEX

Spanish Cedar, 242
Spathodea campanulata, 82–84
Sprekelia formosissima, 48–51
Spurge Family, 151–161
Stemmadenia, 59–60; *palmeri*, 59–60; *tomentosa*, 59
Stenocereus, 121–126; *marginatus*, 121–122; *thurberi*, 121, 123–124; *weberi*, 121, 123–126
Sterculia Family, 337–341
Sterculiaceae, 337–341
Strangler Fig, 253
Strelitzia Family, 342–343
Strelitziaceae, 342–343
Strelitzia reginae, 256, 342–343
Sunflower Family, 132–144
Swietenia humilis, 245–246

Tabachín, 191, 197
Tabebuia, 82, 85–89; *chrysantha*, 6, 85–86; *palmeri*, 6, 85, 87; *rosea*, 88–89
Talauma mexicana, 232
Tar Bush, 135
Tatache, 118
Taxodiaceae, 342–344
Taxodium Family, 342–344
Taxodium mucronatum, 342–344
Tecoma stans, 88–91
Tepozánn, 227
Theobroma cacao, 340–341
Thevetia, 61–63; *ovata*, 61–62; *peruviana*, 61; *thevetioides*, 61, 63
Texas Ranger, 316
Theophrasta Family, 345–346
Theophrastaceae, 345–346
Tigridia pavonia, 178–181
Tiliaceae, 346–347
Tithonia fruticosum, 142–144
Toché, 169
Tochomitl, 175
Toronja, 330
Torote Verde, 170

Tree Cholla, 113
Tree Fern Family, 149–150
Tree Morning-glory, 145
Tree Sunflower, 142
Trompeta, 248
Trompetilla, 304
Trompetilla grande, 77
Trompillo, 99
Tronador, 66
Tulipan de Africa, 82
Tuna, 113

Ulmaceae, 348
Uña de Gato, 203
Uña del Diablo, 239
Uva de la Mar, 299
Uva de la Playa, 299

Vauquelinia corymbosa, 304–305
Velo de Vivuda, 349
Verbena Family 349–355
Verbenaceae, 349–355
Vinorana, 203
Violeta de Agua, 300

Washingtonia robusta, 274–275
Water Hyacinth, 300, 302
Waterleaf Family, 178–179
White Mahogany, 82
White Mangrove, 132
Wigandia urens, 178–179

Yellow Bird-of-Paradise, 192
Yoyote, 61
Yucca, 41–46; *carnerosana*, 6, 42–43; *decipiens*, 44, 46; *filifera*, 44–45; *jaliscana*, 44, 46; *periculosa*, 44, 46; *rigida*, 42–44; *torreyi*, 42, 44; *treculeana*, 42, 44

Zoapatli, 138
Zygophyllaceae, 355–357